# GHOST MOUNTAINS
## AND
# VANISHED OCEANS

# GHOST MOUNTAINS AND VANISHED OCEANS

## NORTH AMERICA
### FROM BIRTH TO
#### MIDDLE AGE

JOHN WILSON & RON CLOWES

KEY PORTER BOOKS

For Arthur Holmes and Ken Walton—J.W.

For Sheila, my wife, for her love and support—R.C.

Library and Archives Canada Cataloguing in Publication

Wilson, John C. (John Charles)
    Ghost mountains and vanishing oceans : North America from birth to middle age / John Wilson and Ron Clowes.

ISBN 978-1-55470-047-9

    1. Geology—North America.  2. Geophysics—North America.  3. Lithoprobe Project.  I. Clowes, Ron  II. Title.

QE71.W553 2009                        557                        C2008-902369-2

The Canada Council  Le Conseil des Arts
FOR THE ARTS  DU CANADA
SINCE 1957  DEPUIS 1957

ONTARIO ARTS COUNCIL
CONSEIL DES ARTS DE L'ONTARIO

The publisher gratefully acknowledges the support of the Canada Council for the Arts and the Ontario Arts Council for its publishing program. We acknowledge the support of the Government of Ontario through the Ontario Media Development Corporation's Ontario Book Initiative.

We acknowledge the financial support of the Government of Canada through the Book Publishing Industry Development Program (BPIDP) for our publishing activities.

Key Porter Books Limited
Six Adelaide Street East, Tenth Floor
Toronto, Ontario
Canada M5C 1H6

www.keyporter.com

Illustrations by Doug Hunter: pages 31, 40, 59, 64, 65, 66, 72, 74, 76, 109, 126, 180, 182, 202, and 208. Illustrations by Philip Hammer, Department of Ocean Sciences, University of British Columbia: pages 29, 104, 105, 124, 130, 133, 136, 140, 142, 155, 161, 220, and 222.

Text design and formatting: Alison Carr
Printed and bound in Canada

09 10 11 12 13 5 4 3 2 1

# CONTENTS

# FOREWORD
## by Bob MacDonald

Don't you hate it when everything you know is wrong?

Take the Earth for example. Good old terra firma: solid, immovable . . . wrong!

The ground beneath our feet is not as firm as it seems. It moves, it tears, it bleeds, it changes shape over time. The Earth is alive and it's taken a lot of hard-fought science to realize that.

It took two thousand years of looking at the skies to figure out that the apparently flat Earth is actually a large ball that's doing pirouettes around the sun and falling through an unimaginably large expanding universe. Then only within the last hundred years did we realize that solid rock is not so solid after all. Continents move. The very face of the Earth changes over time ever so gradually the way your face has changed since you were a baby. A time traveller flying over the Earth during the age of the dinosaurs would have found navigating difficult because current maps would be useless. Back then the continents were all clustered together in one big landmass with one big ocean on the other side of the world

Today we have a new perception of a dynamic Earth that moves and changes. But this view has not come about easily. After all, the idea of an evolving spinning ball in a vast empty cosmos goes against all of our natural senses. Go take a look for yourself. The ground doesn't feel like its moving, the horizon is flat, and mountains that were there yesterday are still there today. But this is our feeble human perception of our planet. We don't live long enough to see how it changes. Our view of the Earth is like an insect that you swat during a summer picnic. The insect has no sense of the four seasons because its entire adult lifetime happens during one summer. So the bug doesn't see how the lush foliage it flies through on a warm day transforms into a frigid landscape of leafless trees surrounded by a white blanket of snow. In the same way, a human lifetime is too short to see the slow steady changes taking place in the body of the Earth.

That human shortsightedness has been at the source of hot scientific debate about the true nature of our planet for more than two thousand years. Aristotle believed the Earth orbited around the sun, but his ideas were eclipsed by Ptolemy and other ancient Greek thinkers who could not imagine a moving Earth. It had to be at the centre of everything. Copernicus revived the idea of a sun-centred universe fifteen hundred years later and Galileo proved it with telescopic observation. But he was chastised by the church for such sacrilege. It took more careful observation, more advances in mathematics by Kepler and Newton before the true motion of the Earth as a sphere tumbling through space was finally accepted.

The same skepticism worked against Alfred Wegener at the beginning of the last century when he proposed the idea that the face of the Earth changes over time, that continents can move. What a preposterous idea. How can solid rock move through solid rock?

Wegener was simply discussing a map of the world with his wife one evening and noticed, as anyone who looks at a map can see, how the east coast of South America and the west coast of Africa have the same shape, as though they were once joined together. It looked to him that some kind of tear in the crust of the Earth had appeared millions of years ago and the two continents slowly drifted apart like ice sheets during spring thaw. But geologists at the time couldn't imagine rock moving through rock and spoke up loudly against this silly idea of continental drift. But Wegener persisted, others took up the idea and now, the theory of plate tectonics is the foundation of geological science. It turns out the surface of the Earth is cracked into about a dozen continent-sized pieces or plates that are pushed around by hot material boiling up from the interior of the planet.

Part of the proof of moving continents came from Canadian scientist Dr. J. Tuzo Wilson who demonstrated a mechanism for moving large pieces of the Earths crust around. He was studying the Hawaiian Islands and noticed that they form a chain of volcanoes with the most active one, Big Island, at one end and the others getting progressively older as you move down the chain. He showed that there is a hot spot under Hawaii where molten material from below burns a hole through the crust forming a volcano. But then as the floor of the Pacific Ocean moves, the volcano is pushed away from the hot spot, goes extinct, and a new hole in the crust is punched through behind it. Hence the chain of Hawaiian volcanoes and proof that the surface of the Earth moves.

Canadian scientists have always been at the forefront of geology. Allan Hildebrand, now at the University of Calgary, discovered a huge hidden crater in Mexico that was formed sixty-five million years ago when an asteroid hit the Earth and gave the dinosaurs a really bad day.

It makes sense that Canada would cultivate world-class geologists. Within our vast boundaries you'll find an example of every type of geology on the planet. From the oldest rocks in the world, the Canadian Shield, to the youthful Rocky Mountains, and the fossil-bearing badlands, it's a geological candy store.

Now Canadian science has taken us one step further, proving once again that our human perception of the ground beneath our feet is only a small sample of the truth. Lithoprobe literally re-drew the map of Canada. Rather than looking at what we see on the surface, it probed one hundred kilometres below ground and revealed a complex landscape that doesn't look anything like we see from above. The laser-flat prairies actually cover the roots of a mountain range that once stood taller than the Rockies. Huge folds in the crust, cracks, faults: The history of how our land was torn apart from Europe and squeezed together into the continent we know today is laid out before us for the first time in exquisite detail.

That's the joy of looking at the world through scientific eyes: It changes everything. To our five senses, the Earth is generally flat. Walk in any direction long enough and you will come to an ocean, so we must be living on a big island. That's not a bad model of the Earth because that's the way it looks. A geologist, on the other hand, sees our country this way: We do live on an island, an island of rock. It is moving toward the west about as fast as your fingernails grow. People on the east coast are waving goodbye to Europe as the Atlantic Ocean gets wider. People on the west coast are waving hello to Asia as our plate heads over the Pacific. But this isn't without resistance. Our moving continent is running over the floor of the Pacific Ocean, and the collision is pushing up the Rocky Mountains. So the Rockies are the crumpled fender of North America. This slow-motion collision is also responsible for all the earthquakes and volcanic activity along the west coast.

While all these rumblings in the ground might seem like a recipe for disaster, we wouldn't be here without them. The gases and water vapour gushing out of all the cracks and holes in the Earth's crust have provided the air we breathe and water we drink. Without all of that geological activity, the atmosphere would burn off into space, oceans would freeze, and the Earth would become an icy desert like Mars is today.

So now we have a pretty clear image of our planet inside and out. But the work is not over. Despite this scientific perception of our planet as a dynamic place, no one can predict exactly when or where the next earthquake will take place or which volcano will blow its top. It's in our best interest to understand the ground beneath our feet. Our lives depend on it.

# PREFACE

When we compare the present life of man with that time of which
we have no knowledge, it seems to me like the swift flight of
a lone sparrow through the banqueting-hall where you sit in the
winter months... This sparrow flies swiftly in through one door of
the hall, and out through another... Similarly, man appears on
earth for a little while, but we know nothing of what went on
before this life, and what follows.

—Venerable Bede 673–735, *History of the
English Church and People*, trans. L. Sherley-Price

W HEN I WAS AN UNDERGRADUATE geology student at St. Andrews
University in Scotland in the early 1970s, one of my professors was given
a piece of moon rock to study. He had been selected because he was an
expert on olivine, a mineral common both on the Moon and in the volcanic rocks of Greenland, where he worked. A local news crew filmed him
proudly holding the specimen up and saying what an honour it was to be
one of the few scientists in the world selected to be given a piece to study.

The sample itself was not particularly dramatic. It was a small piece
of rock that had been glued to a glass microscope slide, then ground

down to a thickness of thirty microns. A micron is a millionth of a metre, so you could stack thirty of my professor's thin sections of moon rock on top of each other and still slide them through a slot one millimetre high. The piece was thin enough for light to pass through—but it was still a piece of the Moon.

I have no idea what conclusions my professor came to about the olivine he had been given, but that's not the point. That tiny sample was part of something incredibly vast—the solar system, the universe, space—it was all my *Star Trek–*, *Outer Limits–*, and *Twilight Zone–*fuelled childhood science fiction fantasies coming true. It was a piece of rock from another world. True, meteorites are from outer space, but this modest thin section was different—*one of us* had gone out there to get it.

Suddenly, the terrestrial rock samples I was studying in my lab classes seemed awfully pedestrian. Not in and of themselves—they too had stories to tell and, to me, many of those stories were far more interesting than the one told by olivine—but they came from beneath my feet. The lunar rock came from the realm of the stars above my head.

Today, moon rock is not that uncommon. The mineralogy sections of many museums have some, and we may soon get more. NASA has plans to return to the Moon, and we are looking farther afield. Rovers on Mars have told us a lot about its geological history, and it is within the realm of possibility that humans will visit the red planet in the not-too-distant future. Remote-controlled spacecraft are studying, or on their way to study, everything in our tiny corner of the galaxy: from asteroids and comets to Mercury, Jupiter's moons and poor old recently downgraded Pluto.

It's all fascinating and I still get the same thrill I got from peering at my professor's moon rock when I look at a beautiful picture of the edge of a crater on Mars, or the latest high-resolution radar images of methane lakes on Titan. It's incredible that we can do all this, but we mustn't lose sight of the one planet that is much easier to study than the Moon or Mars—Earth.

There is a mild irony in the fact that, while I was gawking at the moon rock, and the exploration of space was firing my imagination, and forever altering the way we look at ourselves and our home, there was a fundamental revolution occurring in the much more down-to-earth science I was studying. That revolution is a large part of the story in the pages to come, and it led directly to Lithoprobe, a two-decade-long multidisciplinary study that has added significantly to our understanding of what went on billions of years ago, what is going on kilometres beneath our feet, and, even more extraordinarily, how we ourselves originated.

Lithoprobe was an extraordinary project and this book is, in part, about that project—but this is also a book about wonder. Groundbreaking, world-leading, ahead of its time—all these accolades apply, but Lithoprobe did so much more than exceptional science.

Lithoprobe discovered stories. Stories of mountains rising and falling, oceans opening and closing, and entire continents sliding around on the surface of our planet like shuffleboard pieces. Stories that incorporate lost explorers, Irish bishops, Scottish cannibals, and a naturalist who should be as famous as Charles Darwin. The tale told is of the birth, death, and rebirth of entire continents and, by extension, it is the story of the creation of the world on which we live. It is also *our* story, because the processes that formed our world created and formed us just as surely. And that's where the wonder comes in. It is the same wonder I felt looking at a piece of moon rock, and that I still feel when I find a four-hundred-million-year-old fossil locked in its stone prison. It is the wonder of the past, of being a tiny speck in a nanosecond of time and yet still being able to ask, "Where do we come from?"

Lithoprobe has been as huge a milestone in the study of our home planet as having a Rover trundle about on the surface has been to the study of Mars. It began in a university seminar room in January 1981, in Ottawa, when the outside temperature was considerably lower than the temperature at the surface on the equator on a sunny summer day on Mars.

On that chilly day, one hundred university scientists met to discuss what was wrong with the study of the Earth Sciences in Canada. Why was Canada, home to some of the best geologists and geology on the planet, falling behind in understanding the fundamental structure of the continent on which we lived? The answer lay in the third dimension.

By the time of that Ottawa meeting, much of Canada's surface geology was well understood by the scientific community. A geologist could take you almost anywhere in the country and tell a fairly cogent tale based on what you could see around you. You could pick up a rock and ask, "What's this?" and get an answer in a lot more detail than you probably wanted. You could say, "How did the landscape we're walking over form?" and be amazed by conjured images of kilometre-thick ice sheets grinding past. However, if you were to ask, "What's going on twenty kilometres below our feet?" the answer would have been sketchier.

Canada is rich in oil, gas, minerals, diamonds, and scenery, but even the deepest mines and drill holes only scratch the surface, a surface that has mostly formed over only the past twelve thousand years since the last ice age. World-class scientists from Halifax to Victoria have refined, and are continuing to refine, our view of Canada's geology. They struggle with the eternal research problems of a lack of funding and an overall unifying research perspective but, by the time of the Ottawa meeting, something else was missing. No one was looking deep down. What was below the surface? The two hundred kilometres of the continental lithosphere, including the forty kilometres of Earth's crust, tell a different story than looking just on the surface. It's a complementary story, but it is a much longer and older story than we've been told just by our studies of Earth's surface.

Scientists in the United States and elsewhere were looking at that deeper story. They were looking down, exploiting new technologies to begin to probe the lithosphere itself. Canada was falling behind. That was the crisis in Canada's Earth Science research, the issue that the Ottawa group pondered. The answer they came up with was that Canada needed a project that would combine geological (the direct study of the rocks) and geophysical (the study of rocks using remote sensing) techniques to determine the third dimension of crustal geology. It was the start of a new Earth Science approach in Canada.

Many more meetings followed. In Calgary, four months later and many degrees warmer, the group unanimously agreed that funding should be sought and proposals solicited for one or more major new research projects. These new projects should involve as many disciplines as applicable and should combine the efforts of university, government, and industry.

Meanwhile, scientists, managers, and officials from the Geological Survey of Canada (GSC) of the Department of Energy, Mines and Resources (now Natural Resources Canada) met for a few days "to discuss the Earth Science issues facing the nation today and in the future and the role that the Department should play."[1] One of the ideas floated was the "Deep Earth" initiative, "to provide a sophisticated framework of crustal structure and evolution that may be used as a basis for natural resource exploration and evaluation."[2] The ball was rolling, and a framework was already in place.

The International Lithosphere Program was established in 1980 "to elucidate the nature, dynamics, origin, and evolution of the lithosphere through international multidisciplinary geoscience research projects and coordinating committees."[3] In response, Canada created a multidisciplinary committee of scientists who developed a plan—Lithoprobe—a project to solve key geological problems in specific corridors across Canada. At the follow-up meeting in Calgary, Lithoprobe was considered as one of the major new research projects, and in 1982, Lithoprobe was selected by the multidisciplinary committee as one of the major projects that should be put forward by the Earth Science community for funding from government organizations and industry.

Many fine words had been spoken and a steering committee for Lithoprobe formed, but over the next two years no firm decisions were made. Finally, the GSC authorized half a million dollars to carry out a study of what was going on at depth below Vancouver Island. They threw out a challenge to the university community to come up with equal funds for the survey and to provide support for academics to do the work.

The selection of the West Coast for the survey was a stroke of genius.

This was not going to be an esoteric investigation of a complex problem that could only be fully understood by people with a Ph.D. in geophysics or geochemistry. This was to be a survey over one of the most seismically active places on Earth, the future site of a huge earthquake that would significantly impact the lives of millions of people. The challenge was met by a group of dedicated scientists from a number of universities in western Canada and the GSC. Phase I of Lithoprobe was underway. It was a magnificent success: For the first time ever, the boundary between tectonic plates that generate large magnitude earthquakes was imaged, leading to further studies and much better understanding of earthquake generation and risk in the region.

Of course, even after the science got started, more discussions took place, meetings were held, and proposals for the main scientific studies were solicited and evaluated, all with the aim of establishing Lithoprobe as a national Earth Science research project. A Phase II proposal was developed and sent to the GSC and to the Natural Sciences and Engineering Research Council (NSERC), the Canadian government's primary funding agency for university research. Building on the dramatic success of Phase I, the Phase II proposal was well received.

Over the next two decades, other proposals followed (Phases III to V). There were ups and downs, but success followed success. By the time of Lithoprobe's conclusion in 2005, twenty-two years of outstanding science had produced a great body of knowledge. It was one of the largest and greatest scientific endeavours ever undertaken in Canada. On that cold Ottawa day in 1981, Canadian Earth Sciences left the starting blocks and began the journey that would put them firmly ahead of the rest of the world in the study of Earth, deep below our feet.

—John Wilson

# PART I

# INTRODUCTION

## CHAPTER I

# A VAST JIGSAW PUZZLE

...a Venetian who is a very good mariner and has considerable
skill in discovering new islands...has found two new very large and
fertile islands, and also discovered the Seven Cities.

—Raimondo di Soncina, Letter to the
Duke of Milan, August 24, 1497

Was it for this the clay grew tall?
—O what made fatuous sunbeams toil
To break earth's sleep at all?

—Wilfred Owen, "Futility"

SOMETIME ON JUNE 24, 1497, a handful of Bristol seamen, two
hopeful merchants, a Burgundian, a Genoese barber/surgeon, and an
Italian adventurer, huddled on their twenty-four-metre ship the *Matthew*,
spotted a rugged coastline through the summer fogs of the western
Atlantic. The coast was part of Labrador, Cape Breton, or Newfoundland,
but the mariners thought it was Cathay. They celebrated their arrival and
spent the following weeks searching for cities with roofs of gold and the
eastern source of the silks, spices, and precious stones so treasured by
Europeans. They sailed the coast, going ashore at least once to get fresh

water and erect a cross claiming the land for Henry VII of England. They named all they saw, caught fish in wicker baskets, and, although never venturing inland farther than the range of a crossbow, collected animal snares and a net-making needle, and noted notched tree trunks as evidence of human occupation. After a month, Giovanni Caboto and his men sailed home with no gold or spices but a hold full of wondrous stories and a request that the king finance a return visit the following year with an even grander expedition.

*John Cabot and some of his men in the finery that Henry VII bought for them.*

Henry VII, the first of the Tudor dynasty, was notoriously tightfisted with the royal purse strings, but he was also insecure. His claim to the English throne was tenuous at best. He had won the crown in battle, but much of his reign was spent defending it against pretenders who swarmed into his country from Ireland, Scotland, and France. The last and most serious of them, Perkin Warbeck, arrived in Cornwall shortly after Caboto returned with his tales of Cathay. Warbeck's revolt was soon put down and the leader thrown into the Tower of London. Feeling relatively secure after having defeated all comers in the first twelve years of his reign, and having established his descent from King Arthur, at least to his own satisfaction, Henry was casting about for projects that would enhance his power, prestige, and wealth. Cabot's stories offered him just the chance he was looking for.

*Cabot named his ship, the Matthew, for his wife, Mattea. This replica, built in 1994, now offers short cruises from Bristol harbour in England.*

Like many before and after, Henry decided that exploration in the name of the king was a good political move. It held out the hope of a potentially vast return on investment, and it was cheaper than a war with France. The parsimonious Henry positively threw money at Caboto, who spent the winter dressed in silk as a self-styled Grand Admiral. Even the Burgundian and the Genoese barber became counts. In the spring of 1498, with five ships, enough convicts to establish a colony, and the "...hope to establish a greater depot for spices in London than there is at Alexandria," John Cabot, as he was known in England, sailed again from Bristol—only to vanish mysteriously from the pages of history.

John Cabot was not the first European to visit North America— certainly Norse men had briefly settled, most probably Bristol and Basque fishermen had visited, possibly Irish monks had washed ashore, and perhaps even Romans had straggled along the coast searching for winds to take them home—undoubtedly, the Romans had made it to South America at the time of the Emperor Augustus—but Cabot was the first that we know of to map the coast.

The original map Cabot brought back is lost but, at the time, it was immensely valuable. Not only did it show "New Founde Lands" and support King Henry's claim to them, but it was eagerly examined by the kings of Spain and Portugal to see whether these new territories were part of either of their own slices of the global pie as defined by the Pope in his world-dividing Treaty of Tordesillas of 1494. The map was copied and distributed across Europe and eventually, as people realized that they had not found a new route to Cathay, incorporated into the first rough attempts to define this strange new continent.

Maps are culturally specific. There were maps of parts of North America before Cabot sailed out of the fog. They existed in people's minds, or were scratched on the ground or rocks or hides, and they showed what was needed: trade routes, animal migrations, river crossings. They had no consistent scale—what would be the point of showing details of the long boring trudge to the Caribou crossing place? It was only with the arrival of Europeans and their nascent Renaissance ideas of cataloguing the world and its contents that scale began to be important.

We don't know what Cabot's map looked like, but, given the surveying instruments of the time, it was undoubtedly wildly inaccurate. However, the process it began, the modern mapping of the North American continent, has continued for over five hundred years. Hudson, Cartier, Thompson, Cook, Franklin and a host of others struggled, suffered and sometimes died, just to add to Cabot's few scrawled lines. And the process has got us now to the point where the map is incredibly detailed—with a GPS system, we can locate ourselves within a couple of metres, and we can confirm our maps by looking at the entire continent from space. All the boring bits are mapped out in just as much detail as the interesting bits. But is it truly complete? Not nearly.

The map you can buy at the local gas station or find online is accurate, but only in two dimensions, and even then, it tends to pay unwarranted attention to the minor surface effects by which we humans set so much store. What of the map of the third dimension—depth? What goes on under the thin layer of dirt we strut self-importantly around on? Well, we've been mapping that for only around two hundred years.

If John Cabot and his Bristol seamen founded North American mapping, the extension of that mapping into the third dimension owes its beginnings to a humble canal surveyor, William Smith, who developed many of his ideas within fifteen kilometres of the dock from which Cabot sailed.

Most of Earth's surface is covered by dirt. It is the debris left by the movement of water, ice, and wind, or it is formed in place by chemicals, plants, and animals. Dirt is the flesh covering Earth's bones. In some places the bones poke through, occasionally in spectacular fashion, as in the Grand Canyon, but mostly they are hidden.

For thousands of years, cultures around the world have realized that the bones of Earth continue deep below the surface, and that by following the rocks downward, valuable things like copper, iron, coal, gold, and precious stones could be found and extracted. In many cases, these early miners probably produced working maps, but they were like the maps of the First Nations in North America, they showed what was locally important but they weren't unified over any great area and, more importantly, they weren't predictive.

All that changed when the booksellers A. Taylor and W. Naylor offered for sale "A Map of Five Miles round the City of Bath, on a scale of one inch and a half to a mile, from an Actual Survey, including all new roads, with Alterations and Improvements to the present time, 1799." William Smith, who had been working on construction of a nearby canal, purchased a copy of Taylor and Naylor's map because he had a revolutionary idea. Smith was an extraordinary man, largely self-taught and with an abiding interest in rocks and the fossils they contained. One thing that he had noted during his surveying work was that exposures of rocks in widely different locations were very similar. He began to wonder if they were, in fact, the same rock and if the outcrops might be joined beneath his feet. They certainly looked the same, with a similar pattern of limestone, shale, sandstone, and coal, but these were rock types common throughout the country. Was it possible to prove that two distinct outcrops were the same rocks and that they ran continuously beneath the dirt of the local farmers' fields?

Smith's opportunity, and one of the defining events in the development of modern geology, came when he was appointed surveyor to the proposed Somerset Coal Canal. Here was a chance not only to examine and survey the rock outcrops in detail, but also to carve nothing less than a precisely measured slice through the countryside.

His work on the canal convinced him that there was a way to distinguish outcrops of rock. Smith knew little of evolution, and Charles Darwin was not to be born for another decade, but he did notice that fossils in widely spaced outcrops of apparently identical rock varied.

Limestone forms under specific environmental conditions. Three separate outcrops of apparently identical limestone may only have in common the conditions of their formation. However, if the creatures that lived in that environment differ, then, because life is continually varying with time, we can say something about the relative ages of the three outcrops of limestone. Smith had no overall pattern of evolutionary change within which to place his fossil populations, but he understood that, if the fossils in two of the three limestone outcrops were identical and the third different, then the first two were probably the same rock and could be joined in a map. Using a combination of varying rock types and the fossils they contained, he could distinguish rocks of different ages and plot them on a map. But how to do it? A different map gave the answer.

In 1798, Smith had noticed a map that had been produced as part of a survey of local farming for the Somerset and Wiltshire Boards of Agriculture, in a copy of the *Somerset County Agricultural Report*. It distinguished soil types, woodlands, fields, and outcrops of rock in various colours. Was it possible, Smith wondered, to show the hidden rocks around the canal he was digging in the same way?

Taylor and Naylor's map in 1799 looked oddly medieval in its circularity with Bath at the centre, and it only covered a radius of about eight kilometres, but it suited Smith's purpose admirably. It was a template that accurately showed many man-made and natural features, but in an uncluttered way that left considerable amounts of white space within which Smith could work.

Smith began by adding surface outcrops of rock to the map. There were only three different types in such a small area—an oolite, a rock made of round grains of limestone; the Lias, a bluish limestone full of fossils; and a marl, a red mixture of clay and calcium carbonate. At each outcrop, Smith had carefully measured how much the beds of rock dipped and in which direction they tended to slant beneath the surrounding cover of soil. He extrapolated these measurements to draw in the likely position of the different rocks beneath the surface. Smith then hand-coloured the rocks in shades still used today, yellow for the oolite, blue for the Lias, and brick red for the marl. The result was a true geological map—the first accurate glimpse of what lay beneath our feet and the first indication that the third dimension was ready to give up its secrets.

Smith was not content to map a few kilometres around Bath. Over the years and at extraordinary personal cost, he travelled around the country, an eccentric character dressed in brown tweeds and a wide-brimmed hat, clutching folders of maps and exclaiming excitedly to anyone who would listen about his latest fascinating discovery. He collected what he saw and plotted what he measured, and the result was a masterpiece.

On August 1, 1815, Smith published sixteen intricately engraved, laboriously coloured sheets that fitted together into a huge map almost nine feet high and more than six feet wide. It was "A Delineation of The Strata of England and Wales with part of Scotland" and it was uncannily accurate and hauntingly beautiful. It was also the first geological map of any nation on Earth.

Like Cabot, Smith started something. Over the following two centuries geologists busily mapped the rocks beneath the dirt. The primary goal remained the same, to map the third dimension, but the tools changed: survey instruments gave way to measurements of gravity and magnetism and hammers were replaced by drills that could retrieve samples from kilometres underground. Geologists changed too. Individual polymaths who could hold the entire science in their heads gave way to stratigraphers, who looked at the rock strata; paleontologists, who studied the fossils; facies modellers, who studied the environment that different rock types formed in; and geophysicists, who studied geological forces. It was the last of these that truly enabled scientists to look tens of kilometres down into Earth's crust, and whose work culminated in the studies conducted by Lithoprobe.

### What Is Lithoprobe About?

Lithoprobe's primary concern was how the pieces of Canada's geological puzzle interact and fit together not only in three dimensions, but also in the fourth: the dimension of time. The project's goal was to gain a basic understanding of the continent on which we live, from which we derive resources,  and which generates natural hazards that we must address. It is also important to obtain new, regional-scale information in various parts of Canada that will prove useful to the mining and petroleum industries—information that these industries could not justify acquiring on their own.

Recognized internationally as the best project in the solid Earth Sciences, Lithoprobe owes its success not only to its outstanding science, but also to *how* it went about doing science.

Lithoprobe developed partnerships with two primary governmental funding agencies: the Natural Sciences and Engineering Research Council (NSERC) and the Geological Survey of Canada (GSC). It was also supported by provincial geological surveys, when work was within their jurisdiction, and mining and petroleum industries, when work was within their areas of interest.

The spirit of collaboration, cooperation, and support among scientists from the many different organizations and disciplines was truly inspiring; they all believed in the grand goals of the project. Lithoprobe was a truly multidisciplinary program within the broad Earth Sciences. The range of studies

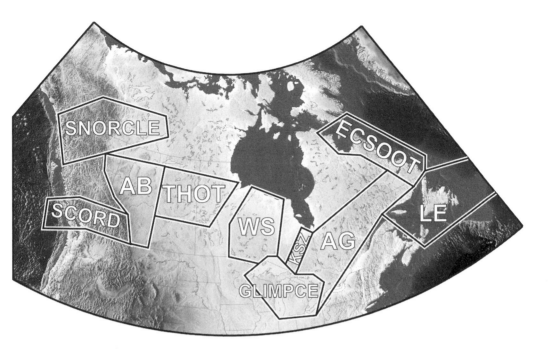

Lithoprobe study areas, or transects, outlined on a topographic map of Canada and the northern United States. From west to east, the transect acronyms are: SCORD—Southern Cordillera; SNORCLE—Slave-Northern Cordillera Lithospheric Evolution; AB—Alberta Basement; THOT—Trans-Hudson Orogen Transect; WS—Western Superior; GLIMPCE—Great Lakes International Multidisciplinary Program on Crustal Evolution; KSZ—Kapuskasing Structure Zone; AG—Abitibi-Grenville; ECSOOT—Eastern Canadian Shield Offshore-Onshore Transect; LE—Lithoprobe East.

contributing to the scientific discoveries was amazing. Virtually every possible approach that was applicable to a problem was used by someone involved with Lithoprobe.

A secretariat at the University of British Columbia provided the centralized coordination and overall leadership, but did not control the scientific programs. During the term of the project, ten different study areas were carefully selected through a national and international review process. These study areas, or "transects," focused on key geological targets that addressed the main theme of the project. Each research transect had one or two leaders and a team of scientists who contributed their expertise to the transect studies. The teams were not exclusive; all scientists who wished to contribute were encouraged to do so. For Lithoprobe research, the old adage "the whole is greater than the sum of its parts" is unequivocally true.

However, looking down is only one part of the equation. William Smith's work around Bath convinced him that the different layers of rock he was looking at were arranged in sequence with the youngest at the top and the oldest at the bottom. This meant that, unless the rocks were turned upside down, the deeper one went, the farther back in time one saw. The maps of the third dimension were also maps of the fourth dimension: time.

Time makes all of geology possible, but geological time is a tough concept to grasp. Geologists and geophysicists work on a time scale that is fundamentally different to the one with which most of us are familiar. There are countless imaginative analogies to show this. My favourite is a version of the grains of rice on a chessboard scenario.

If you are really bored one Sunday afternoon, take a one-kilogram bag of long-grain rice down to the nearest park that has one of those large chessboards marked out on the ground. Starting in the bottom left hand corner, place one grain of rice on the first square. On the second, place two grains, on the third, four grains, on the fourth, eight grains, and so on, doubling the number of grains each time.

There are sixty-four squares on a chessboard and approximately 55,000 grains of rice in your bag, so you, and a number of passersby, will be entertained for some time. But how far will you get? Your bag will be empty before you complete the sixteenth square. That's not even two of the eight rows on the chessboard completed.

So, you were thinking too small. Go and get one of those ten-kilogram sacks of rice that restaurants use. That's about 550,000 grains. It'll be a lot of extra work, but it won't get you much farther. You'll be able to finish off the second row of squares, but even with the kilogram you initially put out, you'll run out again working on only the fourth square along the third row. A bigger bag? A truckload? No, there are not enough grains of rice in the world to complete the sixty-fourth square of your chessboard.

Now, imagine each grain of rice represents one year back in time. A human life span rings in between seven squares (64) and eight squares (128)—almost the complete first row. The first twelve squares encompass

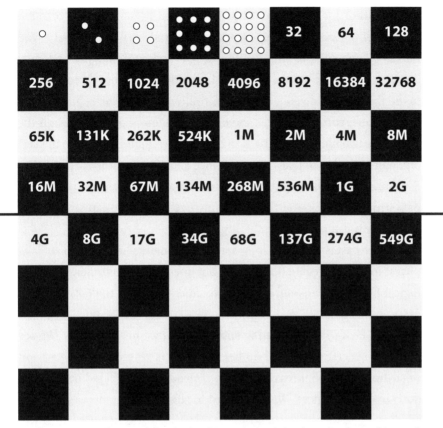

How the years, or grains of rice, overwhelm a chessboard. Dots and numerals show the number of grains of rice on each square of the chess board, starting with one in the upper left, as these numbers are doubled for each square. K, M, and G indicate thousands, millions, and billions, respectively.

human history since Julius Caesar—Christ, Mohammed, Leonardo da Vinci, Milton, Newton, Darwin and Einstein all fit in here. On the fourteenth square when you were only one third of the way through your little bag of rice, you have to include the rise and fall of the Babylonian, Minoan and Assyrian empires. By the end of the second row, as you begin your large bag and are a mere quarter of the way through the board, square sixteen has 32,768 grains/years on it—the Neanderthals are stumbling toward extinction.

Now, as you noticed emptying your ten-kilogram bag, things begin to speed up. The exponential scale quickly snowballs. The third row on the chessboard takes us back to just over eight million years ago, to the geological period called the Late Miocene and early dogs, bears, and horses. Three squares into the fourth row and you are surrounded by dinosaurs, oblivious to their approaching doom. Another three squares, to the thirtieth, and you are back 536 million years at the beginnings of recognizable complex life. At the end of row four, halfway through the board, it's two billion years back in time and a single square into the second half of the board takes you very close to the origin of Earth itself. The Big Bang happened around square thirty-five and it is a dark and unknown past stretching back over the remaining twenty-nine squares, assuming that there was a past before the Big Bang.

So the past is very, very, long and the exponential chessboard analogy puts it in a manageable framework, but does it help us understand geological time in a personal context? No, and it's doubtful whether this is even possible.

Our senses have evolved to show us what we need to know. We see the light spectrum that we do because that is where the information regarding food and threats is. We don't need to see infrared because we don't need to see heat. We don't need to see ultraviolet because, unlike some insects, we don't need to see the intricate patterns on some flower petals. The same applies to our hearing, smelling, tasting, and touching.

We sense the world of our daily experience and do it very well. Hence, we only truly understand things that fit into that range. Someone can explain what nanometres or light years are, but we do not have the capacity to understand it the way we understand a centimetre or a metre. The same applies to a nanosecond or four billion years. So we have to work within something we can understand: chessboards, perhaps, or movies, as in the following analogy.

At times, the movie *Titanic* seems to last forever, but in fact it only lasts 194 minutes, or 11,640 seconds. Let's pretend that the length of the movie is the age of Earth. How would a movie of our world's history look?

The house lights dim and we settle in with our popcorn around 4.7 billion years ago. There is no Earth, no Sun or solar system, just a roiling mass of elements racing away from the cataclysm that began it all billions of years previously. Seven minutes into the movie, as the submersible is finding the safe in the *Titanic*'s encrusted hulk, Earth is coalescing out of the swirling dust. Twenty-one minutes later, as the *Titanic* pulls away from the dock, the oldest rocks yet found on Earth are being formed.

Now there's a slow bit in the movie. For almost the next hour, apart from flurries of meteorite impacts and some impressive volcanic activity, the most important things happening are a slow cooling and the development of simple clusters of cells in a vast ocean. One hour and twenty-three minutes in, and continents are recognizable and the main bits of North America are joining together, just as Jack and Rose (played by Leonardo DiCaprio and Kate Winslet) are kissing on the ghostly prow of *Titanic*. Forty minutes later, the iceberg has done its work and, as what was probably the first supercontinent forms, Rose struggles to release Jack from the flooding lower decks.

As the band plays "Nearer My God to Thee," mountains rise and fall and North America comes very close to splitting apart. Hard-bodied life blossoms in 570-million-year-old oceans and begins the complex struggle for existence as Jack dies and Rose is rescued.

The last two hundred million years of Earth's history, including the rise and fall of the dinosaurs and their replacement by the mammals, all fit into the time it took the *Titanic*'s credits to roll. Even if you are one of the movie fanatics who sit until the very last minute, just to see who the key grip was, you'll still probably be out of your seat and halfway down the aisle before *Australopithecus* raises its head above the grassy, African veldt.

Unlike *Titanic*, this movie of Earth's history will never win any Oscars, because there's far too little human interest, but it puts things in perspective. The planet had a vast history before we came along to think about it but, for all of the remarkable geological discoveries of the past centuries, we still have only a very fragile skeleton of knowledge on which to hang the story of 4.54 billion years. Most of the flesh on the scientific bones of Earth's history is a product of our thinking. Hopefully it is rational thinking that contains glittering gems of truth, but it is still thinking and as such is the product of the culture and perspective of the thinker.

For example, fossil sea shells can be found in many parts of the world and have been interesting to humans for a very long time. The fossil shells are facts, but the interpretation of them has varied. They are the devil's horns, sports of nature created in place within plastic rock, or evidence of the immense power of the biblical flood. I won't even touch on the multitude of guesses about how old they might be.

Which brings us back to geological time. If we cannot comprehend five hundred million years, we cannot think of a process to form fossil shells that requires that much time. The concept of vast time has to be in place before someone can imagine, "maybe these were once real shells that were buried in mud, then sank deep within Earth where pressure and temperature altered the mud to rock before it was all lifted up again and the rock above the shells eroded away grain by grain to leave the shells exposed on a hillside for me to find."

Perhaps the persistence of the idea that Earth is only six thousand years old is a consequence more of our innate inability to comprehend very large things like millions of years, rather than Biblical revelation. For all the chessboard and movie analogies, maybe we need a personal experience of the variety of time to drive it home. Maybe time is personal and, first and foremost, needs to be understood on a personal level. My first glimmerings of an understanding of time, both its scale and its variety, coincided with my discovery of geology.

I register 1964 as an age ago. *Goldfinger* was playing in cinemas. Jean-Paul Sartre won and declined the Nobel Prize for Literature. Martin Luther King, Jr., won the same prize for Peace. The Rolling Stones did not yet look as if they had been freeze-dried, and were taking America by storm. The U.S. Surgeon General affirmed that smoking cigarettes caused cancer. In the fall of that year, I was a terminally shy, confused thirteen-year-old beginning his second year of high school outside Glasgow in the west of Scotland. I suppose there must have been bright spots, but the only one that sticks in my mind is an afternoon spent trying not to throw up in a foul-smelling quarry outside the village of Bridge of Weir.

I had cycled the eight kilometres (oddly, the radius of the circle covered by Smith's original map of Bath) from my home to the quarry because my English teacher had given me a project—I had to stand in front of my twenty-five or so fellow students and talk for five minutes. That five minutes looming before me felt like an eternity. As a self-conscious introvert, I habitually spent most of my energy on making myself invisible— blending in was safe, and standing before the chalkboard was not.

To say I was unhappy at the prospect of my upcoming speech would be a gross understatement. I was petrified. I lost sleep, lying wide-eyed and struggling to think of something I could talk about for such a vast length of time. I wrote speeches and read them out, only to discover with horror that they lasted thirty seconds and left four and a half minutes for the class to laugh at me. I considered running away to sea or feigning leprosy, but I sensed that nothing would work. They would catch me and stand me up to speak anyway.

The day loomed so large in my imagination that I could see nothing past it. Life was coming to an end and that seemed a shame so early, with so much yet to experience. Then Alex, who lived in Bridge of Weir and was also something of an outcast because of his geographic inability to participate in the after-school tribal rituals, told me about a place where there were sea shells in the rock. I was convinced he was lying to me, trying to get me to believe an absurdity so that I could be ridiculed, but he seemed so genuine that I got directions to this mysterious place to check it out.

That Sunday I cycled out and found it. The quarry was not large but it was old, dating back to the fifteenth century when local monks had dug coal there. More recently, a neighbouring tannery had used it to dump waste and it was half full of thick, black, utterly disgusting sludge.

The smell hit me as soon as I dumped my bike at the crest of the ridge at the quarry entrance. It was rotten eggs, and so powerful that I gagged.

What I should have done at that point was get on my bike, pedal home and stumble through a presentation on my pet Scotch terrier, Meg. I didn't for two reasons. First, the odour didn't scare me as it should have. It was familiar from the stink bombs that were occasionally let off in the school corridors and I didn't know it was dangerous. It was only many years later that I learned the smell belonged to hydrogen sulphide, a byproduct of the chemicals used to break down the hair on animal hides, and discovered that hydrogen sulphide is comparable in toxicity to hydrogen cyanide, and that "even a low level of exposure to the gas induces headaches and nausea, as well as possible damage to the eye. At higher levels, death can rapidly set in and countless deaths attributable to the buildup of sulphide in sewage systems have been recorded."[4]

The second reason that I didn't get back on my bike and race home was a small piece of ribbed rock that stuck out of a boulder beside my left foot. I set to with my father's claw hammer and, after a few minutes work, held a fragment of what was obviously a sea shell in my hand.

That was it! I was sold. The smell vanished from my consciousness, and I spent the rest of the day scrambling around the edge of the sludge, uncovering impossible wonders from the quarry walls. My prize was a complete bivalve, rather like a fat clam, that I dropped and had to immerse my arm up to the elbow in tannery waste to retrieve. I cycled home, happy and smelly.

In the short term, I learned the value of good props in a presentation. My classmates were as fascinated as I had been and, after a few introductory sentences, they were so engrossed in the treasures I passed around that they completely forgot to humiliate me.

I also learned something about time. It was flexible—five minutes could be an eternity or a flash—and it was big. The monks who mined the coal had gone to meet their Maker a long time ago, but the shells and other creatures I dug from the rock that afternoon had been dead immeasurably longer. They had been born, procreated, and died in the muddy seabed of the Visean Stage of the Lower Carboniferous—345 million years before I stumbled around the quarry (that's between the twenty-ninth and thirtieth squares on the chessboard, if you're keeping track).

In my spare time, I became as eccentric as William Smith had been, dedicating my Sundays to cycling around the west of Scotland armed with a hammer, hand lens, newspapers, notebook, and tiny bottle of 10 per cent hydrochloric acid (the acid fizzes when dropped on limestone) and returning late in the evening laden with bags of carefully labelled rocks. I read everything I could lay my hands on and I went on to do a degree in geology at St. Andrews University. It was mostly fun, too much fun, occasionally, and my fossil collection grew steadily. Then I got a job.

It was in Africa remapping part of something called a Greenstone Belt in Zimbabwe. My first day was a bit like a return to the horrors of preparing my five-minute presentation. I went to a river section where the air photos indicated that there was good exposure of the rocks to be mapped. There was, but it didn't help me much. I was obviously surrounded by several distinct rock types. They had different characteristics and were various shades of dark green, but *what* they were, was a mystery. There wasn't a fossil in sight.

That morning the true scale of geological time hit home. In dealing with fossils, I had blithely talked about three, four, even five hundred million years, anything younger than the Cambrian Era, which had begun around 570 million years ago. These green rocks were more than four times that old, 2.9 billion years, and I was still little more than halfway back to the beginning of Earth. I had been interested in the grains of rice/years on the first thirty squares of the chess board but squares thirty-one to thirty-three contained the other seven-eighths of Earth's history.

Squares thirty-one to thirty-three represent the Precambrian Era, almost the first three hours of the movie, the time before life with any kind of shell or skeleton raised a head or antennae to look around, and it's a time that is often ignored, but not by everybody.

A geologist friend of mine once told me that the only rocks worth working on were older than one billion years—anything younger than that he dismissed as "dirt" and he would have liked little more than to be able to scrape it off to look at the important stuff beneath. Of course, in doing so, he would be losing tens of millions of years of Earth's history, not to mention most of the world's cities and the majority of her population, but he considered it a small price to pay.

Few geologists are that extreme, blithely dismissing thousands of metres of rock as "dirt," although most would agree with William Smith that the top few metres of soil and glacial till that obscure their beloved rocks are a nuisance. Fortunately, Lithoprobe allows us to travel back in time to this strange and distant past, to see beneath the dirt, however it is defined, and paint the beginnings of a picture of how our world has formed and changed.

---

### Seismic Reflection Method

For many decades, the oil and gas exploration industry has used the seismic reflection method as the principal means of imaging the subsurface. Nothing short of drilling gives the details of rock layers and structures as does this method. Over the years the technology has changed and improved a great deal, much due to computer-based developments. The scientists of Lithoprobe adapted the seismic reflection technique to look into the lithosphere at depths of fifty kilometres or more. Due to the high costs of a seismic reflection survey, Lithoprobe has always run two-dimensional profiles along existing roadways, rather than cutting paths through the bush or running three-dimensional surveys as is common in petroleum exploration.

The principles of the seismic reflection method are pretty basic. In essence, we are carrying out a simple ultrasound of the Earth below us. And just as a doctor needs to interpret the results of a medical ultrasound, seismologists need to interpret the fuzzy images that they produce and relate them to the surface geology and other information that has been acquired.

A source of sound (i.e., seismic) energy is initiated at the surface and produces seismic waves that propagate deep into the subsurface. All of Lithoprobe's land-based surveys used the "vibroseis" method: four large (twenty-five-ton) trucks, mechanically and hydraulically equipped with pounding

Lithoprobe

*A Dancing Elephant at rest.*

pads, vibrate the ground synchronously for about twenty seconds, and do this up to eight times in a row to get more energy into the ground. At the boundaries between different rock layers, at structures such as faults and even throughout the crust at small-scale heterogeneities, some of the source energy reflects back toward the surface; the rest continues propagating down. At the surface, a long array of 480 individual sub-arrays about fifty metres apart extends over twenty-four kilometres (twelve kilometres to either side of the source). Each sub-array records the merged signals from nine to twelve sensitive electro-mechanical devices called "geophones." About five thousand geophones are placed in the ground to "listen" to the reflected energy for up to forty seconds and transmit the signal to a computer. Using its recording of the four trucks' new twenty-second signal vibration, the computer processes forty seconds of "listen" time into a new twenty-second signal that appears as if it were generated by a pulse of energy. The eight sets of new signals are then added together. This represents one "shot," a term that harkens back to when this was done with explosives. The vibrator trucks move along the line one hundred metres and repeat the process, with the array moving along with the sources. Reflection surveys that are hundreds of kilometres in length and more than fifty kilometres in depth are recorded in this way. Because the depths are greater than the lengths to one side of the source, the method is also called the near-vertical reflection technique.

After the survey is completed, computers in a large processing centre further process the recorded data to produce a two-dimensional image of the subsurface below the road along which the data were acquired. Lithoprobe recorded about twelve thousand kilometres of land-based seismic reflection data during its twenty-year history.

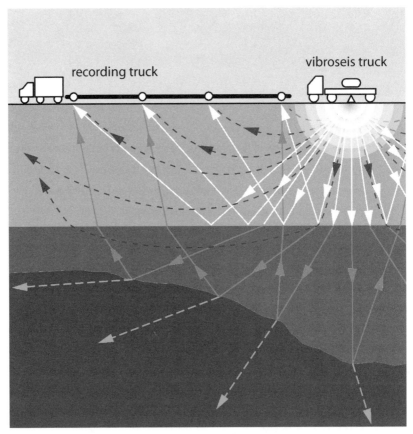

recording truck

vibroseis truck

Schematic representation of a seismic reflection survey. Solid lines show the paths of reflected waves whereas the dashed lines show the paths of refracted waves (see "Seismic refraction method"). The distance from the source (vibroseis truck) to the furthest recording device is about 12 km for a reflection survey, whereas the depths from which reflections can be obtained exceeds fifty kilometres; hence the terminology "near-vertical incidence reflection".

*A "Texan" refraction recorder connected by cable to the ground motion sensor, or geophone, which has a spike so it can be planted in the ground. The recorder is about twenty centimetres long.*

*Texan recorders and geophones being readied for deployment during an experiment in Utah. In many refraction experiments, Canadian and U.S. seismologists worked together and shared instruments.*

### Seismic Refraction Method

The seismic refraction method, a complementary approach to seismic reflection, is only used by academic and government scientists; it doesn't provide the resolution that the exploration industry requires. Scientists, however, often look at the large-scale regional picture, rather than searching for a specific structure that may contain oil or gas, and the complementary information provides us with information about the types of rocks that exist at depth and an indication of the temperatures at those depths. Generally, such studies are done in the cheaper two-dimensional mode along existing roadways, although some three-dimensional studies have been carried out.

The principles of seismic refraction are very basic and revert back to earlier studies of light refraction by a prism. In seismic refraction, sound waves are refracted by the varying rock types below Earth's surface. Sound energy travels in different rock types at different speeds. Generally, rock types with greater sound speeds are farther below the surface. Sound energy that is initiated at or near the surface travels downward in hemispherical waves. These waves continually "refract" or bend toward regions that have rocks with slower speeds. At some point, the sound waves bend so much that they start coming back up toward the surface. There, they are recorded by sensitive seismographs that consist of a geophone or seismometer (technically a little different than those used in the seismic reflection method), a recorder and a device for precise timing such as a global positioning system (GPS) unit.

Where there are major changes in rock types below the surface, reflections can also be generated. We call these "wide-angle reflections" because generally they are recorded at distances from the source that are much greater than the depths to which the sound waves have propagated. Reverting back to the analogy with light waves, at some particular distance away from the source, most of these sound waves are "totally reflected" at the major boundaries. This simply means that the angle at which the rays associated with the sound waves impinge upon the boundary at those distances is such that all of the incident energy is reflected back upwards to be recorded at the surface. Have you ever gazed at a calm mountain lake and seen the perfect reflection of the mountains in the water? The image you see is caused by total reflection of the light from the mountain by the surface of the lake. We get the same thing with sound waves, but the image is much more difficult to sort out.

How do we actually carry out a seismic refraction/wide-angle reflection experiment? First, we need a large source of sound energy. On land, this usually means an explosive charge. We drill holes in selected locations and load these holes with chemical explosives, usually from a few hundred to a thousand kilograms. The sound waves must be able to be recorded from near the source to many hundreds of kilometres away because the greater the distance from the source, the deeper the refracted energy has penetrated. For example, to reach Earth's crust at a depth of about forty kilometres, we need to record signals out to three hundred kilometres from the source. For greater distances, say up to a thousand kilometres, and thus greater depth penetration, say to about one hundred kilometres, we need larger charges. Then we use multiple holes near one another, each loaded with a thousand kilograms. On these long profiles, we place the explosive sources about fifty kilometres apart. Individual seismographs are placed along the roadways at one-kilometre intervals. Thus, a seven-hundred-kilometre profile would involve about fifteen shots and deployment of seven hundred seismographs. It is a big operation.

The times at which the sound waves arrive at the instruments and their amplitudes are used in computer models that emulate the wave propagation. After testing hundreds of models, a cross section of the sound-speed variation along the recorded profile to depths of fifty kilometres or more is generated. We can relate different rock speeds to different types of rocks to help determine the composition of the crust. Also, high temperatures tend to lower the speed with which sound propagates, so we can estimate crustal temperatures at some depths.

---

But why? Is there any point to all this? If we can only truly understand a human scale of minutes, hours, days, and years, what's the point of looking back unimaginably far to a time when nothing, not even the chemistry of the atmosphere, would have been recognizable? The point is the same as the one that makes studying our primate ancestors important—to know where we come from.

An understanding of the geology of Earth is essential to truly understand our place on it. To put it the other way round, we cannot understand life and our place in it without understanding the ball of rock on which, and out of which, it evolved. We are not just passengers on a dead piece of cosmic debris whirling through space; we are an integral part of an exceptional, dynamic system that produced both ourselves and our Earth. In a very real sense, geology made us.

Not that this was particularly difficult. Amino acids are the basis of life, and they are everywhere: in volcanic craters, around hot vents deep beneath the ocean and in dust and meteorites from space. In fact, you could make some in your kitchen: take a mixture of carbon, hydrogen, nitrogen, and oxygen, and add energy—heat, microwave radiation, etc.— and, within a day or two, you will probably end up with amino acids. The next stage is tougher and you will need a bottle of the gases that tend to burp out of a volcano. But when you get them, bubble them slowly through your solution of amino acids at room temperature. This will give you bonded combinations of amino acids, or "peptides." A chain of peptides is called a "polypeptide" and a chain of polypeptides is a "protein." Proteins are the building blocks of cells, which constitute life: you and me.

To make it even easier, although not in your kitchen, bombard this organic soup you've made with lumps of muddy ice that are rich in oxygen, nitrogen, more amino acids, and amphiphiles. These icy lumps are comets and they cause the organic soup and the amphiphiles to react with water to form membranes, essential for making cells and keeping all the other neat stuff together.

Of course it's not easy, but our infant Earth had all the elements, the energy and, most importantly, the geological time necessary for this experiment to be carried out on a vast scale, probably many times. Given the composition of early Earth and the energy available from its core and the Sun, simple cellular life was probably inevitable. Where we were fortunate was in our Earth's size and distance from the Sun.

Our near-twin planet, Venus, is much the same size as us but much closer to the Sun. It has also been tectonically active with a molten core and huge upwellings and downwellings of magma in its interior, but its surface, due to a runaway "greenhouse" effect, has no water and is extremely hostile to life. Surface temperatures range up to 475°C, the atmospheric pressure is ninety times the pressure on Earth, and the atmosphere is mainly carbon dioxide.

Mars, on the other hand, is much farther from the Sun and considerably smaller than Earth. Its small size meant that it cooled quickly, becoming an inert lump of rock about a billion years ago. Its current atmosphere is primarily carbon dioxide with a few droplets of water vapour and its pressure is only around one percent of that on Earth. The planet's surface water is locked in polar ice caps. Did life forms once exist there? Maybe. The jury is still out, but many of the necessary elements were available to Mars, once. Any organism that managed to put itself together on Mars would have been extremely simple and the conditions would not have encouraged evolution into anything more complex.

In contrast, life on Earth kept going, the dynamic core of our planet pumping up heat and chemicals to encourage and nurture the biological experiments burbling on its friendly surface.

Darwin caused a sensation by suggesting that we had evolved from ape-like ancestors. Now we accept much more radical ideas than that. In fact, we are descended from small rat-like creatures that scurried between the legs of dinosaurs; from fishes, struggling to drag themselves from mud hole to mud hole on proto-limbs; and from odd, flat worms with a backbone so primitive it is barely recognizable. Take it farther back, and we are descended from clumps of organized cells floating and reproducing in a planet-wide ocean. Or, to say it even more extremely, we are descended from proteins, polypeptides, and amino acids—inanimate elements coming together in volcanic craters or around deep sea sulphide vents. As the poet Wilfred Owen suggested, we are "clay grown tall." Our Earth made us. We are the children of volcanoes, and it behooves us to remember our origins.

In looking at Earth's origins and its development, we are looking at our own origins and the development of a home that allowed us to survive and encouraged us to thrive. And there are few places better to look at our Earth than North America.

North America, including Greenland, Central America, and the

Caribbean islands, is the third largest continent on the surface of the globe. It includes the world's largest island, second and fourth largest countries, second biggest lake, third lowest point, fourth longest river, and the eleventh and twelfth most extensive deserts. Its rocks contain a staggering wealth of oil, gas, coal, diamonds, gold, precious and base metals, and some of the world's best places for digging up dinosaur bones. It is bounded by two mountain ranges—the Rockies to the west and the Appalachians to the east—and is home to more than five hundred million people. North American rocks range from Earth's oldest to youngest. Unfortunately, William Smith's discovery that younger rocks are piled on top of older ones means that many of the really old rocks are buried under immense thicknesses of younger layers. Fortunately, there is Lithoprobe.

## A Legacy for Canada

More than a thousand scientists contributed to Lithoprobe, approximately half of whom were students or postdoctoral fellows carrying out their training in a collaborative, multidisciplinary environment. Over its twenty-two-year span, the project generated more than fifteen hundred peer-reviewed publications and articles, and greatly expanded our knowledge and understanding of the four-dimensional geology of Canada. An outreach program developed at the Lithoprobe Secretariat brought exciting and informative results to the public through the media; to educators and students in schools, colleges and universities; and to other targeted groups such as politicians and funding agencies.

The energy sector also benefited from the program's regional studies, which provided an enhanced knowledge base from which the industry's own more detailed exploration and development plans could be prepared. A regional study in the Western Canada Sedimentary Basin where most of the country's oil and gas resources are concentrated exemplified this benefit.

In a variety of mining locations associated with base metals, diamonds and uranium throughout Canada, Lithoprobe's studies provide a valuable framework of knowledge and understanding that otherwise would not exist. Scientists applied the high-resolution seismic reflection technique to mineral exploration problems in mining regions where expensive infrastructure was already in place. A number of companies then proceeded with their own seismic programs, including the use of three-dimensional procedures. The Geological Survey of Canada's involvement in Lithoprobe spawned a major new geophysical program to detect ore bodies by combining knowledge of seismic, electromagnetic and rock properties techniques to mining camps.

Lithoprobe scientists with the GSC designed and built a new magnetotelluric (MT) instrument including the necessary software. Such a device detects and records minute changes in the electrical and magnetic fields induced in Earth from naturally occurring currents of charged particles in the stratosphere, particles that originally derived from the Sun. Interpretation of MT data enables identification of rocks that have high electrical conductivities, such as ore deposits or, at much greater depths, rocks that might be partially molten, from normal rocks that have much lower conductivities. The GSC instrument was a special long-period version that allows scientists to look hundreds of kilometres deep into Earth. The technologies were transferred to a Canadian company, Phoenix Geophysics, which specializes in MT work. The firm continued its research and development on the instrument in cooperation with the GSC until recently. Phoenix sells the instruments worldwide and carries out contract surveys using them. In their field, they are the pre-eminent company in the world, partly through their association with Lithoprobe.

---

The many studies that made up Lithoprobe looked at questions as diverse as when and how the first continents formed more than four billion years ago, and why North America's west coast will one day suffer a devastating earthquake. Lithoprobe significantly contributed to the story of our Earth, how it was formed, and how it has evolved and continues to do so. The project did not answer all the questions, but it did add important bones to the skeleton of our understanding. A major unifying theory that tries to explain how that skeleton moves and, in passing, helps us understand our own place in the scheme, is plate tectonics.

# The Beckoning Cross
## by Barrie Clarke, Dalhousie University

IN LATE JUNE OF 1994, I was at home in Halifax watching a movie on
TV with my teenage daughter. The movie told the story of a brother and
sister who embarked on some important mission, the details of which I
have forgotten, but early in their adventure the brother was killed in an
accident. So important was their mission that the boy's spirit continued
to carry it out to a successful conclusion with the sister. In the closing
scene of the movie, the sister took her brother to the cemetery and
showed him a headstone with his name on it. Only then did he realize
that he had died in the accident.

Two weeks later, I was carrying out my own Lithoprobe mission in
uninhabited northern Saskatchewan. I was with my assistant, Andy
Henry. We were doing shoreline geology and  obtaining rock samples
from south Davin Lake, to understand the age, origin, and evolution of
the different granite rocks and what they could tell us about tectonic
development in this region. As we came around a point, we noticed a
white cross on the lake shore at the end of the bay. In thirty years of
doing fieldwork all over the world, I had never seen a white cross in the
woods, so we hastened to the end of the bay to find out what it was all
about. When we arrived at the cross, I was in the bow of the boat and able
to read its inscription, but Andy was back in the stern and could not. He
asked, "Barrie, what does it say?" I struggled to break the silence that
had come over me. Finally, I blurted, "It has my name on it!"

Now, I regard myself as a person who is always strongly connected to
reality and is always able to maintain his equilibrium, but this was the most
unsettling, disorienting, surrealistic moment in my life. Memories of the
recently watched movie immediately flooded back. For an instant, I wondered if Andy had brought me to this place to tell me that I had died. How

Photo by Barrie Clarke

*The mysterious cross on the shores of Davin Lake, Saskatchewan.*

long it took me to re-establish my composure, I don't know. Eventual closer inspection of the other side of the cross indicated that *this* Barry Clarke was only fifteen years old when he died in 1987. Even so, for the next few weeks isolated in the field, I kept wondering—if the probability of finding a memorial cross in the woods is infinitely small, what is the probability that if you do find one, it has your name on it? What did it all mean?

When Andy and I eventually returned to civilization, we made inquiries about Barry Clarke. We learned that he was a Woodland Cree First Nations boy who had died in a hunting accident, and that his home had been in Southend, Saskatchewan. When I returned to Halifax, I wrote a short letter addressed to "The Clarke Family Who Lost a Son on Davin Lake in 1987, Southend, Saskatchewan," telling them of my discovery and enclosing some dried flowers in his memory. A week later, my phone rang and it was the boy's mother. The Southend Clarke family was as overwhelmed as I was, particularly Barry's sister Pauline who had carried the package home from the post office and couldn't believe that the name of the sender in its upper left hand corner was that of her brother. Over the following years, we exchanged many letters, but never had an opportunity to meet each other. However, en route to the GAC-MAC meeting in Saskatoon in 2002, I arranged to stop in Regina to meet Pauline, who had since become a student at the University of Regina. We had a wonderful brunch and afternoon together, and later that summer, I met the entire Clarke family back at Davin Lake on what would have been Barry's thirtieth birthday. We spent most of the day together, and as we sat around the campfire that night, I felt as comfortable as if I had known the Southend Clarke family my whole life. This unparalleled "Clarke Family Reunion" was filmed by CTV/ATPN. A short version appeared on *CTV News with Lloyd Robertson*, and the small documentary has been shown on *Indigenous Circle* on APTN.

Pauline has the best interpretation of the meaning I had been searching for to explain my discovery on Davin Lake. She believes that Barry's spirit brought me from Halifax to find his cross. Her explanation of this connection between us makes much more sense to me than any sterile finite probabilities. She takes comfort that Barry's spirit operates through me, and as a result she calls me "Bro." In return, I refer to her as my "Spiritual Sister." I've tried my whole life to get my friends to spell my name correctly as "Barrie," without much success, but now I sign my email messages to Pauline as "Barry." Thanks to Lithoprobe, no two other people on Earth have the same relationship as we do.

## CHAPTER 2

# FLOATING CONTINENTS

On a round ball
A workman that hath copies by, can lay
An Europe, Afrique and an Asia,
And quickly make that, which is nothing, All.
—"Valediction: Of Weeping," John Donne

It may seem easy, said Piglet, but it's not everyone can do it.
—*The House at Pooh Corner*, A. A. Milne

Aɴʏ ꜰɪᴠᴇ-ʏᴇᴀʀ-ᴏʟᴅ ᴄʜɪʟᴅ who has done a *Sesame Street* jigsaw puzzle will notice on a map of the world that South America and Africa fit together as well as Oscar the Grouch fits in his garbage can. The child who notices this unknowingly follows in the footsteps of Francis Bacon. In 1620, when the natural sciences were in their infancy, Bacon commented on the curious similarity of the two Atlantic coastlines. Between Bacon and Oscar, many others have pondered this similarity. The trick, however, has always been to explain the apparent congruency.

In the hundreds of years after Bacon, most attempts to understand the puzzle involved vertical rather than lateral movement. Continents subsided and became oceans or rose and became dry land. It seemed to make sense; after all, didn't this fit with the biblical stories of the flood and and/or Plato's description of the fate of Atlantis? The few who suggested lateral movement of the continents were, for centuries, ignored or ridiculed.

But evidence that the continents were once much closer together was growing. The pictures on the faces of the jigsaw pieces seemed to fit. The patterns of the coal-bearing rocks of Europe and North America and the fossils in them matched. Thick sequences of sediments on New Zealand appeared to have washed off Australia. Gold deposits in Ghana could be explained as deriving from source rocks in Brazil, Guyana, and French Guiana. All this can be seen to make sense if the continents fit together in the ancient past and then moved apart. Even the lowly and much-maligned eel has something to say on the matter.

Eels are slimy and difficult to skin (the best way is to nail their head to a post or tree, slit around just below the gills with a very sharp knife, and peel the skin off like a glove), but their flesh tastes wonderful when lightly fried in olive oil with garlic. They also have very odd mating habits.

Atlantic eels fall into two species: the North American eel (*Anguilla rostrata*) and the European eel (*Anguilla anguilla*). Each species lives in the rivers and estuaries of its appropriate continent, but both return to the Sargasso Sea to breed. The Sargasso Sea may have been mentioned  by the fourth-century poet Rufus Festus Avienus, and is the only sea in the world without a shore. It is a huge area of the Atlantic Ocean (1,100 km wide by 3,200 km long) of warm, seaweed-clogged water lying between the Azores and the West Indies, and latitudes 20 and 35 north. These are known as the "horse latitudes" because so many of the Spanish and Portuguese ships sailing through them found themselves becalmed and had to throw their horses overboard to conserve water. The surface of the sea is covered with sargassum seaweed and the area has an evil reputation as the graveyard of ships. However, at certain times of the year, the waters below the surface host a writhing, joyful copulation of eels.

The mystery of the Sargasso Sea eels is in trying to explain how two very similar species can breed in the same place and yet unfailingly divide into two distinct populations that undertake very different journeys to opposite locations to live out their lives as adults.

The North American eels have by far the easier time of it. Their journey back to the eastern seaboard is short, although it takes the young glass eels a year to make it. Tough though that journey is, however, it is nothing compared to the odyssey of the European glass eel that faces a two- to three-year, almost six-thousand-kilometre journey back across the Atlantic.

Two different populations of eels, breeding in the same place, but undertaking vastly different journeys back home. Clearly, something has happened since the Atlantic eels' ancestors squeezed through the closing gap between North and South America some fifty million years ago.

Perhaps the ancestral eels once faced a similar trek regardless of whether they headed east or west out of the Sargasso Sea. Perhaps the ancestral Hudson and Severn Rivers were equally accessible. Unfortunately, at least for *Anguilla anguilla*, eels never developed the science of geology. Unbeknownst to the eels, as the eel offspring wriggled and drifted their way homeward, things were going on far beneath them. The Atlantic Ocean got wider, and each year the eastward-headed eels had a centimetre or two more to travel. For countless generations it made no difference. However, eventually, the changing distances meant that the two populations were arriving to breed at slightly different times. Given enough time, two separate species arose, one blithely continuing its age-old journey totally unaware that its close cousins were struggling with an ever-more-daunting migration.

Could moving continents, therefore, give rise to different species? Quite possibly, but no one has proved it. This is hardly surprising considering the many mysteries that still surround eels—no one knows how they navigate such prodigious distances over open ocean, nor has anyone ever observed them mating beneath the sargassum. The eels' odd habits do not prove that continents move, but less speculative biological evidence comes from the other side of the globe. It grew from the efforts of an extraordinary man whom chance placed in exactly the right spot to make some amazing discoveries.

The naturalist Alfred Russel Wallace owes what fame he has to having come up with the theory of evolution at the same time as Charles Darwin. He did it in four years of zealous collecting in the Malay Archipelago, much faster and more intuitively than the plodding Darwin. While encountering headhunters, battling snakes and leeches, and suffering unremitting dysentery and recurrent bouts of malaria, Wallace collected hundreds of reptiles and mammals, thousands of birds, butterflies and insects, and tens of thousands of beetles. This quarter of a million specimens led him to two dramatic epiphanies.

The island of Ternate is a mere one hundred square kilometres of volcanic rock jutting out of the Ceram Sea. It is one of the smallest of the 17,508 islands that comprise modern Indonesia, but its history belies its apparent insignificance. Ternate is one of the fabled Moluccas, the Spice Islands over which, in the days when cloves could fetch a higher price than gold, the European nations fought savagely for generations. In 1579, Francis Drake dropped anchor there during his circumnavigation. He presented Sultan Babu with a magnificent velvet cloak, concluded a trade treaty, and left. His ship was so weighed down with cloves that the *Golden Hind* was, according to Drake's report, "laid up fast upon a desperate shoal." Rather than part with the cloves, Drake's men threw their cannons and considerable quantities of food overboard before they refloated their vessel.

As it passed through Portuguese, English, and Dutch control, Ternate was a major source of valuable spices. However, its lasting fame, and the reason why it should probably be a shrine to modern science, lies with what happened in a remote hut there in 1858.

Wallace was recovering from malaria when, as he later wrote, "There suddenly flashed upon me the idea of the survival of the fittest." His fevered mind had discovered a mechanism for the appearance of new species—individuals varied, and less-well-adapted ones tended to die, while better-adapted ones were more likely to live longer and reproduce. With enough time, the positive variations accumulated to create a new species. It was a defining moment in science and in the birth of our own world view.

Wallace dashed off a letter and a short scientific paper and sent them to Darwin. It came, Darwin said, like a "bolt from the blue." This was exactly the idea Darwin needed to tie together all his long years of study and research. In a remarkable case of scientific cooperation, Wallace's paper and one by Darwin were read together at a meeting of the Linnean Society on July 1, 1858. Prodded by Wallace's discoveries, Darwin published *On the Origin of Species* the following year and a new species of human thought began its long, and sometimes troubled, evolution.

Wallace's second great idea was read to the Linnean Society on November 3, 1859 (the paper's author was still busily collecting and thinking in the Far East). He had noticed that his vast collection of creatures fell into two broad groups: those that were familiar in Asia (apes, monkeys, tigers, elephants, bears, squirrels, deer, etc.) and those from Australia (kangaroos, duck-billed platypuses, opossums, flightless birds, cockatoos, wombats, and so on). These two populations were as different as the inhabitants of Africa and South America, but instead of being separated by thousands of miles of water, the line dividing them bisected narrow ocean straits, sometimes only twenty-five kilometres wide.

The line Wallace drew (now named after him) marked where these two radically different animal populations came together but it marked, Wallace realized, something even more extraordinary: a fundamental break in the crust of Earth.

He knew nothing of the breeding habits of eels, but he did know that the animal populations of Africa and South America were different. If these differences had arisen through divergent evolution as the two continents slowly moved apart, what would happen if the land masses came back together? In his remote jungle hut, he understood that this must have been what happened along his line. In mapping animal diversity, he was also mapping the edges of two continents, Asia and

Australia. It was a glimpse of the opposite process to the divergence of Africa and South America, and it could not be explained merely by subsidence and flooding.

Alfred Wallace's remarkable intuition did not take him the next step and explain the processes that created his line, but his observations helped another Alfred put it all together and prove to a doubting world that the continents really had moved.

Alfred Wegener was a generalist, interested in almost every aspect of the world around him. He spent years in the Arctic where he observed two ice crystal halos that are now named after him; he studied the weather and helped discover why raindrops are the shape they are; and he wrote a revolutionary book.

*Alfred Wegener pondering the fit of the continents.*

While looking at a map of the world, Wegener noticed what Bacon had seen three centuries before, that Africa and South America appeared to fit together. But Wegener went a step farther. What if *all* the continents had once been together? What if it could be proved?

Almost obsessively, Wegener worked his way through any science that might produce evidence to support his idea. He looked at mountain ranges that were cut in half by oceans, rock patterns that fitted into a bigger picture if the continents were moved together, and fossils that were the same across huge distances. He also looked at evidence of ice

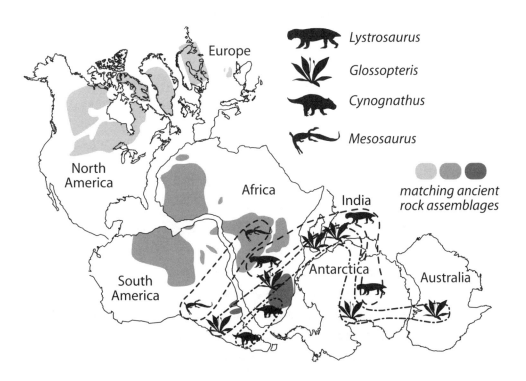

The jigsaw-puzzle fit of the shapes of continents formed the basis of Alfred Wegener's theory of continental drift, but much additional evidence supports his theory. When the continents have been placed together, we see that similar ancient rock types are found in North America, Greenland, Europe, South America, and Africa. Evidence from paleontology in specific areas ties South America, Africa, Madagascar, India, Antarctica, and Australia together. Mesosaurus, a Permian (ca. 260 million years ago) freshwater reptile, is found in central South America and Africa. Lystrosaurus, a Triassic (ca. 230 million years ago) mammal-like land reptile, is found in Africa, Madagascar, India, and Antarctica. Cynognathus, an older Triassic reptile, is found in Argentina and southern Africa. Glossopteris, a fossil fern, is found within the indicated zone in all of the southern continents.

ages that made sense only if the continental bits were reassembled around an ancient pole. He studied geology, climatology, biology, zoology, and paleontology, and he examined Wallace's Line.

It all made sense. Once, Wegener theorized, there had been a single supercontinent, Pangaea. It had broken into two smaller continents, Laurasia in the north and Gondwanaland in the south. Eventually these also broke apart: Laurasia to give us North America, Greenland, Europe, and most of Asia; Gondwanaland to produce South America, Africa, Arabia, India, Australia, and Antarctica. Wallace's Line was where a piece of ancient Laurasia and a piece of Gondwanaland were coming back together after hundreds of millions of years of wandering.

In 1915, Wegener published his work in a book, *The Origin of the Continents and Oceans*. In it he proved that the continents drifted around the surface of Earth. Unfortunately, he had no idea how. Because of this, his ideas were dismissed by much of the scientific community of his day. He was ostracized and found himself unable to obtain a university position in his native Germany. But Wegener was an extraordinary man. Content that his work was sound, he ignored the sometimes vicious attacks on his book and continued studying his beloved natural world.

In November 1930, when Wegener was aged only fifty, he and a companion set out for the coast from a weather observation camp in Greenland. It was bitterly cold, but both were experienced Arctic travellers. Unfortunately, this time, something went horribly wrong.

The following May, Wegener's body, a smile on his face and his eyes still open, was found by another scientific party in a sleeping bag on a glacier, dead of apparent heart failure. His companion was never located. The discoverers erected a huge iron cross as a memorial and departed. But the glacier moved, carrying Wegener and the cross with it. By the 1950s there was no sign of either. The delightful irony in this is that, while the esteemed scientific hierarchy ignored the man who had got the moving jigsaw puzzle of Earth's crust right, his body was slowly being carried away by ice, in a human-scale parallel of continental drift.

It is in the nature of scientific inquiry that radical ideas are often

ignored until the traditional old guard dies off and a new generation
looks at the problem in a different light. This was the case with Wegener's
idea. In the 1950s and 1960s, three different areas of study, initially
completely unrelated to continental drift, came together to provide the
missing link. With Wegener's ideas, these forged an unassailable unifying theory of global geology that revolutionized the way Earth scientists
think and determined much of Lithoprobe's future work.

When molten lava cools, certain iron-rich minerals within it become
magnetized by Earth's field. They orient themselves north-south in the
solidifying slurry and become fixed as the rock hardens—tiny, fossilized
compasses. These compasses point not to the true geographic north and
south poles that are fixed in our imagination and which Frederick Cook,
Robert Peary, Roald Amundsen, and Robert Falcon Scott struggled so
hard to reach, but to the magnetic poles, the points where the lines defining Earth's magnetic field converge at the top and bottom of the planet.
They are easier to reach than the geographic poles, but they are entirely
more slippery characters to define.

Early Arctic explorers rapidly discovered that their compasses were
more or less useless at high latitudes. Instead of reassuringly pointing
north, they began to indicate west or east or somewhere between. The
compass needle also exhibited a disturbing tendency to point down. It
was this last feature, the dip of the compass needle, that led to the discovery of the magnetic poles. When the compass needle is pointing
straight down, you must be at one of the points where the lines of magnetic force are curved back down into Earth's magnet.

Using this characteristic, the British explorer Sir James Clark Ross
found and visited the North Magnetic Pole in 1831. Referring to the
medieval idea that the magnetic poles were marked by mountains of
magnetic iron, he expressed a slight regret that "nature had here erected
no monument to denote the spot which she had chosen as the centre of
one of her great and dark powers."[5] Ten years later, Ross became the first
man to visit the South Magnetic Pole or at least get very close to it. On
this occasion he regretted not reaching the precise spot as he had long

cherished the hope of "being permitted to plant the flag of my country on both the magnetic poles of our globe."[6]

Actually, Ross knew that flag planting at the magnetic poles was a futile exercise. Even while taking his measurements in 1831, he had noticed that the pole was moving. Later explorers had to chase the pole as it wandered across the Arctic. The North Magnetic Pole has moved about 1,100 kilometres since Ross visited it, and it appears to be speeding up, whipping along at forty kilometres a year. This is enough to place it in Siberia by 2060.

If you carefully orient and study the tiny compasses within samples of lava of different ages and the same location, you can plot the historical course of the magnetic poles, on a "polar wandering curve." Alternatively, if you measure the orientation of the lava compasses in rocks of the same age, but from different localities, they should all point to the same place, the location of the magnetic pole when the lava cooled. This is exactly what was done in the 1960s with samples of lava of the same age from all around the Arctic. The results were surprising.

Instead of pointing to the same spot on Earth's surface, the magnetic north pole at the time the lava solidified, they pointed all over the place, to multiple north poles. There cannot be more than one north pole, so the rocks themselves had to have moved away from a position where all the micro-compasses once pointed to a single pole. The magnetic minerals in the lava were recording the breakup of Laurasia.

This is supported by plotting the polar wandering curves for North America and Europe. The curves are the same shape, but they are offset and only match if the two continents are moved side by side.

## Magnetism and Continental Drift

When a lava flow or a rock that has been heated cools to below 500°c, the magnetic compasses are "frozen" and provide the direction of the magnetic field at the time this occurs. This direction comprises two parts. In a horizontal plane, the compasses point the direction to the magnetic poles. In a vertical plane, the compasses typically show a dip away from the horizontal and this dip provides the latitude. At the Equator, the dip is zero; at the magnetic poles, the dip is a 90° rotation. Between the equator and the pole, the dip varies with latitude.

One of the early discoveries from studies of ancient magnetism (or paleomagnetism) was a demonstration that the magnetic poles of Earth reversed themselves at varying intervals of time. This discovery goes back to the early 1960s. An Australian graduate student doing field work in the outback found an ancient campsite that had been inhabited by Aborigines. They cooked their meals in a fire pit that was made of stones, and he noted that the stones were magnetized. He carefully removed some of the stones to study in the laboratory, but before doing so he sketched and recorded their physical setting, including orientation. Taking the stones back to the laboratory, he carefully analyzed the magnetization in these rocks and found that it was exactly opposite to that which would be caused by Earth's present magnetic field. Knowing that the campsite was about thirty thousand years old, he reported to his supervising professor that at that time, Earth's magnetic field must have been the reverse of what it is today. The professor was incredulous, but further studies proved the student to be correct.

During the 1960s, studies of paleomagnetism blossomed. And when paleomagnetism was combined with dating of the same rocks, it was possible to determine an absolute time frame for the magnetic reversals. At the same time, studies such as dredging samples and extensive mapping of the sea floor were underway and provided remarkable results. On the basis of this new information, a Princeton University professor, Harry Hess, proposed that new ocean crust was forming at the recently discovered ocean ridges due to upwelling of molten rock from the mantle below. The new crust is carried away from the ridge in a conveyor belt fashion as the uprising mantle material spreads laterally.

Some of the sea-floor mapping included measurements of the intensity of magnetism using magnetometers that had been developed during World War II. When these measurements were plotted, they showed zebra-striped patterns that became critical information leading to acceptance in the late 1960s of the sea-floor spreading hypothesis.

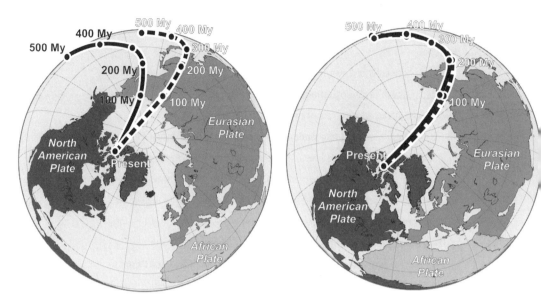

The left globe shows simplified versions of the apparent polar wander paths for North America (solid curve) and Eurasia (dashed curve) for the past five hundred million years. The paths are similar in shape but different in location on the globe. They diverge because North America has drifted westward by about twenty-four degrees with respect to Eurasia. The right globe shows the positions of the apparent polar wander paths when the land masses are reassembled. The divergence is eliminated. There is only one north magnetic pole.

Dated reversals of the magnetic field, upwelling of molten material beneath ocean ridges, and sea-floor magnetic anomaly stripes all came together in the minds of a few scientists during the 1960s. Their idea was that as molten rock was added to the ocean floor at ocean ridges, it cooled below 500°c and the little magnets in the rock were frozen in the direction of Earth's magnetic field at the time. As time went on, this material spread laterally away from the ridge. Meanwhile, new molten rock was upwelling at the ridge and would acquire a different magnetization if Earth's field had reversed. This is the sea-floor spreading hypothesis: the ocean floor is one gigantic tape recorder.

Much controversy attended the hypothesis, both in its proposal stage (ca. 1963) and through to its acceptance stage (ca. 1968). A Canadian, Larry Morley of the Geological Survey of Canada, suffered most from the controversy. In February 1963, he submitted a letter paper to the prestigious journal *Nature* in which he effectively postulated the sea-floor spreading hypothesis. After a couple of months, the paper was returned with a statement that the journal did not have room to print his letter. So Morley submitted his paper to the *Journal of Geophysical Research* in April. In late September, he received a rejection effectively stating that his hypothesis represented nothing but speculation and could not be considered as serious science. During this same time period, and working totally independently of Morley, Fred Vine and Drum Matthews of Cambridge University in the United Kingdom had a similar revelation about sea-floor spreading. They submitted their paper to *Nature* after Morley's submission had been returned. However, their paper was published in September 1963 and is recognized as the formative paper on sea-floor spreading. To be fair, the Vine and Matthews paper included a lot more

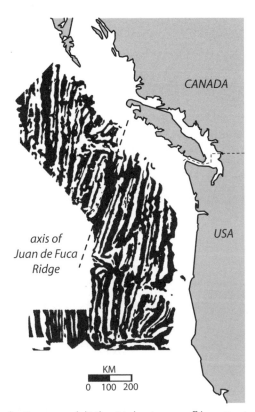

Map showing the zebra-striped magnetic anomaly patterns as recorded in the original marine survey off the west coast of North America in the late 1950s. Black represents normal polarity intervals; rocks in these regions enhance the normal magnetic field, giving high-intensity anomalies. White represents reversed polarity intervals; rocks in these regions reduce the normal field to give low-intensity anomalies.

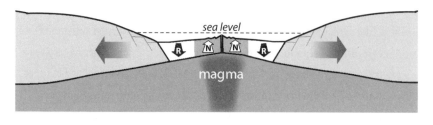

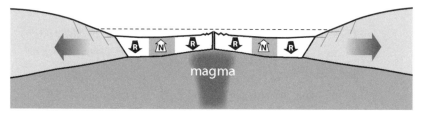

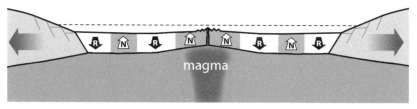

The ocean floor as a gigantic tape recorder. As new basalt (volcanic-like rocks) is added to the ocean floor by the cooling of the partially molten magma, it is magnetized in the direction of the Earth's existing magnetic field. The upper panel illustrates this for a period of normal (N) magnetization, which is the same as that of Earth's present field. At a later time and after a reversal of the magnetic field, the basalt at the ridge is magnetized in the reverse direction (R) and the ocean floor has become larger (middle panel). The process continues (lower panel). The ocean floor behaves like a tape recorder as it registers each reversal of Earth's magnetic field.

scientific data on which their hypothesis was based. Nevertheless, in a letter about Morley's paper published in the journal *Geology* in April 1974, Norman D. Watkins wrote that it is "probably the most significant paper in the Earth Sciences ever to be denied publication."

The second piece of evidence was already in place, waiting to fit into the new theory while the circum-Arctic lava was being measured. In 1955, as an add-on to their project to map the seabed between California and the Queen Charlotte Islands, the American military undertook a study of the magnetism of the oceanic crust. Reams of paper came out of the recording machine as the ship crossed back and forth off the coast, and what gradually appeared was extraordinary. Magnetically, the seabed looked like a zebra, with north-south-oriented black and white stripes alternated in staggering profusion.

The scientists knew what the stripes represented: another odd characteristic of Earth's magnetic field. Not only did the magnetic poles wander, but they frequently reversed themselves, north becoming south and vice versa. On average over the last ten million years, this has happened every 250,000 years, but averages mean little with the fickle poles. The timing of the switches is random, ranging from as little as twenty thousand years to as much as fifty million. For most of the past 375,000 years, the North Pole has been as we now find it. However, even within this short period of time, three reversals of short duration have been identified.

*Why* the poles reverse is a mystery, as is why the reversals themselves should take different amounts of time depending on latitude. For example, the last reversal took two thousand years at the equator and ten thousand years in mid-latitudes. All this proves beyond a shadow of a doubt that the picture I remember from a high-school physics text of Earth as a bar magnet was grossly inaccurate. However, despite knowing little more about the changes in Earth's magnetic field than we do about the mating habits of eels, we can apply our little knowledge of those changes to explain the magnetic zebra-striping of the ocean floor.

Each stripe on the Pacific Ocean floor represents a magnetic reversal, marked by new lava that picked up Earth's polarity at the time that it emerged from the crust and cooled. What was remarkable was that the more the scientists looked at the zebra stripes, the more they began to see a pattern. It was possible to draw a line through the stripes so that the pattern to the east of it was a mirror image of the pattern to the west. There could be only one explanation: new crust was being formed along a specific line on the ocean floor, picking up a magnetic orientation, and then being split into two stripes that were then pushed away from the spreading centre as yet more lava formed. As the signature changed and the lava was pushed outward, they created an identical pattern of magnetic stripes on either side.

This was a major step forward. Wegener had not presented any idea about what happened to Earth's crust as two continents drifted apart. Here was the answer, laid out with elegant simplicity on sheets of recording paper. New crust formed along lines, or ridges, in the middle of new oceans.

But Earth wasn't getting bigger. If new crust formed as the continents drifted apart, where did the excess go? Appropriately enough, the answer to that was found deep beneath where Alfred Wallace had drawn his line.

The strength of gravity on Earth's surface is related to the density of the rocks beneath—the greater the density of rocks beneath any given point on Earth's surface, the stronger the gravity. You won't feel the difference, but it can be measured.

Photo courtesy of William Hueke

*The Mid-Atlantic ridge, an undersea chain of volcanoes where new oceanic crust is being formed and which separates the North American plate from the Eurasian plate.*

## Mapping with Magnetics and Gravity

Magnetometers, which measure magnetic fields, were originally developed during World War II to detect enemy submarines. They were further refined and adapted after the war so that they could be flown from small aircraft to take readings almost anywhere. Many types of rocks contain little grains of magnetic minerals that impart a characteristic magnetic "signature," which can be mapped using magnetometers to show magnetic anomalies. Magnetic anomaly maps are typically generated by flying an airplane along a series of lines and recording the magnetic field values, then removing the values of Earth's normal magnetic field to leave only the "anomalous" values, which are contoured or colour-coded to generate the map. Because the magnetic anomalies are related to the rocks that are used to generate a geological map, a magnetic anomaly map can be used to complete and verify a geological map that is made from limited rock exposures over a given area.

Earth's gravity field can also be used to infer geology, particularly the types of rocks below the surface. Different rocks have different densities. A piece of iron ore is heavier than a similar-sized piece of granite, which is heavier than a similar-sized chunk of sandstone. Geophysicists can use these properties to infer the presence of different types of rocks in the subsurface. To do this, they use gravity anomaly maps. Sensitive instruments called gravimeters measure and record the value of Earth's gravitational pull at a particular location on the planet. All that is needed is an accurate latitude, longitude and elevation. During the late 1990s, instruments were developed to accurately measure gravity from airplanes. Because Earth is a flattened sphere, its field of gravity is different at different latitudes. A gravity measurement also depends on elevation because if the measurement is made up a mountain, the distance to the centre of Earth is greater than when the measurement is made in a valley. A gravity anomaly map is prepared by subtracting values of Earth's normal gravity field, which is determined from a mathematical equation based on Earth's shape as an oblate spheroid, at a particular position and correcting the value as if it were made at sea level plus a few other corrections. The anomaly map then represents differences in density of the materials below the surface. Geophysicists can use this map of differences to try to determine what the densities are and what types of rock cause them.

---

Imagine a large shopping mall. It is early on Tuesday morning, so there are very few people around. However, the coffee shop somewhere in the mall is packed. Assuming an equal distribution of walls, space and fittings around the mall, the mass of people in the coffee shop is an area of greater density than the rest of the mall. If you were walking about on the flat roof of the mall, dropping small steel balls, the balls would accelerate slightly faster over the coffee shop because the greater density of people would have a slightly greater gravitational attraction. You would need very sophisticated equipment, but you could tell where the busy coffee shop was in the mall by doing a gravity survey over the roof and spotting variations in the gravitational pull of what was beneath.

Conversely, if the coffee shop was closed but the mall was packed on the last Saturday before Christmas, you could also map the location of the shop by its lack of density. Using the same principles, you could find masses of dense rock, ore bodies for example, or less dense rock, the roots of vanished mountains, buried deep within Earth.

Earth's crust has a lower density than the mantle beneath it. Therefore, it produces lower gravity readings—a measure of the strength of gravity at any given point on Earth's surface is a rough guide to the thickness of crustal rocks below.

Crustal rocks are thick beneath continents (greater than thirty-five kilometres) and thin (about six kilometres) beneath oceans. However, in the 1920s some surprising results were found beneath the ocean off the Malay Archipelago. A huge gravity low was recorded in a line that corresponded to a deep scar in the ocean bottom. The low-density crust was being dragged down into the denser mantle rock along this line to produce the anomaly. Convection forces in the mantle were moving the crust above, pulling it down into a thick wedge. Remarkably, the calculated speed of this movement exactly matched the speed Wegener had calculated for the breakup of Laurasia and Gondwanaland. This was the

Photo courtesy of Dept. of Physics, University of Toronto

*The Canadian who put it all together, J. Tuzo Wilson.*

process of crustal destruction that balanced the creation of new crust along the mid-oceanic ridges.

All the major pieces were in place. All that was needed were some brilliant scientific minds to tie up a couple of loose ends and see the big picture. One of those, the dominant Canadian in the story, was a University of Toronto geology professor, John Tuzo Wilson.

Wilson contributed three things: he determined that the Hawaiian chain of volcanoes was caused by the crust moving over a hot spot below, he discovered a new kind of geological fault that explained the mechanics of the spreading along the mid-oceanic ridges, and he brilliantly put his findings all together as a key contribution to a new universal theory called plate tectonics.

Suddenly Alfred Wegener was vindicated. Opposition to the idea of a dynamic crust collapsed and it became glaringly obvious how things were happening. Earth's low density crust was divided into a collection of about twelve major rigid segments, or plates, which slid around at a few centimetres a year on top of the denser mantle. New crust was created by volcanic upwelling along ridges beneath the oceans where the plates were moving apart. Old crust was destroyed where it was pulled down, or "subducted," into the mantle where two plates came together. The process was driven by heat moving in immense convection cells, rising through the mantle from Earth's core, cooling and falling back down in exactly the same way as heat circulates through a pan of fudge on a stovetop. Where the heat rose to the surface and spread out, continents were dragged apart and new oceans opened. Where the cooler currents met and fell back, the crust was dragged down with them.

With exquisite simplicity, plate tectonics explained our dynamic

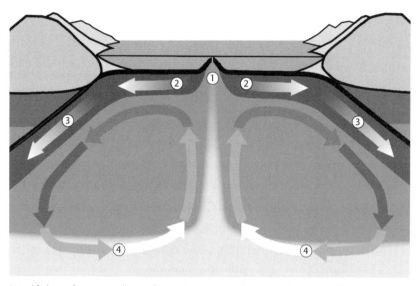

A simplified view of convection cells in Earth's mantle. Hot material from the mantle rises (1), which causes the ocean plates to form and diverge (2). At other locations, plates collide and one cooled plate is dragged below a different one (3). It sinks into the mantle, is heated and rises again at a divergent boundary (4).

world. In detail the processes were unimaginably complex, but as a broad, unifying theory, it was perfect, explaining everything from ancient ice ages and gold deposits to fossil distribution and Wallace's Line. Most significantly, it explained how mountains were formed and why earthquakes occur where they do.

The moving plates could only consist of either thin, relatively new (less than two hundred million years old) oceanic basalts, or thick, old (as much as four billion years old) continental granites, volcanic rocks, and sediments. So there could only be three broad types of collision: ocean-ocean, ocean-continent, and continent-continent. Of course, in detail there was endless variation and the plates need not meet head on. They also can slide past each other without any vertical offset. In California, the North America and Pacific plates slide past each other along the San Andreas Fault, causing the many earthquakes in that region. In Canada, the equivalent fault is the Queen Charlotte Fault off the west coast of the Queen Charlotte Islands and extending northward off the Alaska Panhandle. Canada's largest recorded earthquake, magnitude 8.1, occurred along this fault in 1949; it ruptured a segment of the fault about five hundred kilometres long.

When two ocean plates meet, one is forced down beneath the other and a line of volcanic islands, known as an arc, rises from the water. Examples of volcanic island arcs include the Aleutians, Japan, and the Solomon Islands.

When a piece of continental plate is pushed against an oceanic plate, the denser oceanic plate is overridden and forced deep beneath the continent where it melts and rises to be spewed out on the surface as lava, as from the volcanoes of the Cascade Range of the Pacific Northwest of the United States. The jerky movements along these fundamental joins in Earth's crust are expressed as earthquakes that occur beneath the leading edge of the American plates from Alaska to Chile. As part of the process, the continental edge crumples, faults, and picks up debris sitting on the ocean plate to form a mountain belt, or "orogen." The complex mountain chains of western North America and the Andes of Peru formed this way.

The process of two continents coming together is even more dramatic. The Mediterranean Sea is the last shrinking remnant of an ocean that once separated Africa and Europe. The Alps, the Caucasus, Mount Vesuvius, the earthquakes that still ravage Turkey and Iran, and the

*Where it's all breaking apart, the San Andreas fault in California.*

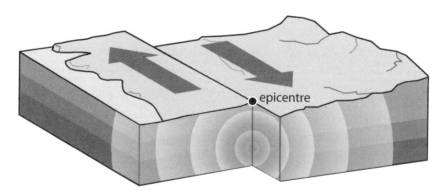

A transform plate boundary occurs when two plates slide past each other with little or no vertical motion, forming a transform (or transcurrent) fault. Such faults can be the source of very large earthquakes, such as those that occur along the San Andreas fault in California.

volcanic cataclysm of Thera and the accompanying earthquake that destroyed the Minoan civilization 3,500 years ago all resulted from this coming together of continents. Similarly, the collision of India and Asia crumpled the crust to form the Himalayas, the highest mountains on Earth—and this collision continues to force the mountains up faster than the patient elements can wear them down.

As soon as the theory of plate tectonics took hold, scientists scurried all over the world to fit Earth's major physical features into this new scheme and, broadly, it worked—oceanic ridges were where they should be and wherever plates came together chains of mountains, or orogenic belts, were forming. And plate tectonics gave birth to even grander unifying theories. But there are difficulties.

One of the first difficulties was that not all the mountain belts were where plate tectonics said they should be. The Appalachians were on the passive edge of North America where no collisions were occurring. Similarly, the Urals were deep within a stable continental block. Did this mean that there was a separate process at work, or were these enigmatic mountains fossilized plate collisions? As it turns out, they were the latter, and mapping them gave rise to a refinement of plate tectonics.

Alfred Wegener proposed a supercontinent, Pangaea, as a precursor to the breakup that has created our current world. But what if Pangaea wasn't the only supercontinent? What if supercontinents have been forming and breaking up over and over throughout the long history of Earth? What if they didn't always break apart and heal back together in the same places?

Recent theories suggest that plate tectonics operates cyclically on a scale of around five hundred million years. The idea is that, as soon as a supercontinent forms, it is doomed. A large crustal mass, such as Pangaea, traps heat rising through the mantle. As the heat builds up, the supercontinent bulges up in the middle, until, after millions of years, the crust cracks and begins to break up. Convection in the mantle drags the broken-off pieces apart and new ocean crust forms between them. Oceanic crust is much thinner than continental crust, and so allows heat

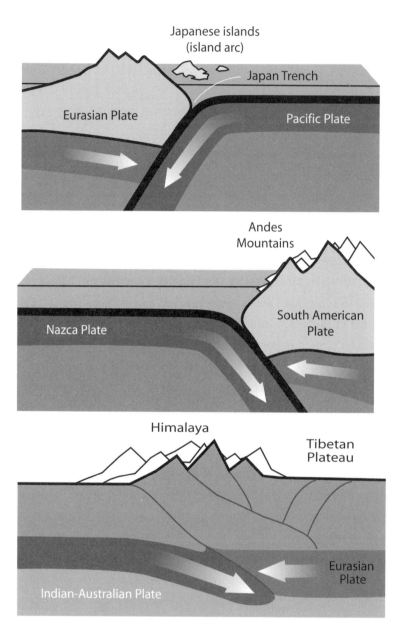

Subduction of an oceanic plate beneath another oceanic plate generates a deep-sea trench and a series of islands called a volcanic arc (top panel). There are many of these islands in the southwest Pacific Ocean. Subduction of an oceanic plate beneath a continental plate forms a mountain belt and crunches and deforms the margin of the continent (middle panel). In a continent-continent collision, the continental crust crumples and thickens, generating high mountains and a wide plateau behind the mountains (bottom panel).

to escape. As the continental fragments separate, more and more heat escapes and the forces dragging the crust around weaken. Eventually, the fragments come back together to form another supercontinent and the heat builds up again.

It's an attractive idea. Like all great theories, plate tectonics and the supercontinent cycles seem almost self-evident as soon as they are proposed. Lithoprobe work has contributed significantly to understanding the shadowy supercontinents that may have existed prior to Pangaea, but there's still a long way to go.

The crust is not beautifully divided into ten or twelve rigid plates with well-defined boundaries. Many more microplates have been postulated in recent years to accommodate anomalies, and their boundaries are sometimes hard to define. Anomalously old ages have been determined for oceanic rocks that should be young. For example, no part of the Atlantic Ocean floor should be older than the last time America and Europe were together, yet some ages much older than that have been found. Spreading ridges at the centre of ocean basins appear to be much more complex than first thought and the zebra-striping of magnetic rocks on either side is not always symmetrical, and occasionally is missing altogether. We still don't completely understand the mechanism that drives the movement of the plates.

These, and other anomalies, don't disprove plate tectonics. However, they do suggest that the simplified diagrams presented as fact in many textbooks, including those shown in the sidebars accompanying the text here, are a gross oversimplification. The history of Earth and the activities of its apparently stable outer shell are immensely complex and, for all our wonderful theories, we are probably only scratching the surface of understanding. We need new techniques and new ideas to fill in the details of the broad general theories. That's where Lithoprobe fits into the evolving story of plate tectonics. It has extended our knowledge back to an ancient time from which only fragments of clues remain, and has increased our understanding of how the processes happening today affect our world.

# Fire!
## by John Wilson

EVERY CANADIAN GEOLOGIST has an armful of bear stories; indeed there's one in this book. In my fieldwork days I had my share of frights walking around a corner in the trail and freezing as two hundred kilograms of *Ursus americanus* lumbered away, huffing in annoyance at being disturbed. However, notwithstanding all the dramatic moments that I elaborate for the entertainment of my thirteen-year-old son, my one moment of true fear was much more long-drawn-out than a chance encounter and had nothing to do with bears.

It was about the time that Lithoprobe was getting off the ground and I was working for the Alberta Geological Survey, collecting and logging abandoned exploration drill cores from the area south of Lake Athabasca. It's a beautiful part of the country dotted with Sahara-like sand dunes and crystal clear lakes. Unfortunately, the summer I was there was remarkably dry and large wildfires had raged across thousands of hectares of northern Saskatchewan and Alberta. I had already lost two irreplaceable drill cores to a fire that had left nothing but a blackened pile of rock.

I had saved the best location for last. It was an old camp by a lake with beautiful sandy beaches. There was a well-constructed hut where my field assistant and I could sleep (it had been built by the drillers as a sauna by the beach), and piles of well-labelled, organized drill cores. There was even a table to lay the cores out on for description. We both thought this was fieldwork heaven and, after the float plane dropped us off, we spent a couple of days working in the shade of some stunted pines and falling into the refreshingly cold waters of the lake whenever it became too hot. The sky was unremittingly blue and the only wildlife we saw were the northern pike that swam over the sandy bottom of the lake to check out the beer bottles we placed there to cool.

It was idyllic and the work went well. On the afternoon of the third day, we radioed for the float plane to pick us up the following morning. At dusk, we sat on the dock (yes, the exploration company had even supplied a dock for us), drank our last two cold beers and watched a lightning storm move along the northern horizon. There was no rain but several bolts of lightning grounded spectacularly in the bush across the lake.

We were just about to turn in when I noticed that, although the storm had moved off, there was still an odd glow in the sky. We sat back down. The glow got brighter, and redder. One of the lightning hits had started a fire. It was a long way off, but there was a wind, and it was blowing toward us.

So how far away *was* the fire? we wondered. I had read somewhere that the horizon at sea, if you are standing on the deck of a smallish ship, is about ten kilometres away. The land was flat to the north and we weren't on a ship, but it seemed like a good enough ballpark figure. The fire was becoming clearer, so maybe it was a good deal closer than ten kilometres, but I'm an optimist.

How fast was the fire moving? We had no idea. It was getting brighter and bigger but that might just be because dark was falling. There was a brisk breeze and the underbrush was tinder dry. I had read stories of fires moving faster than a man could run, but those must be extreme circumstances, so on a pleasant evening such as this, probably no faster than walking speed.

How fast did I walk? I stood up very quickly on the edge of the rickety dock as I remembered calculating hiking times. A brisk walk covered four to five kilometres an hour. Even a slow walk would cover two to three kilometres per hour. The fire would be coming round our little lake by midnight.

Next question: how long could we stand up to our necks in a cold lake with pike nibbling our toes? Long enough, I hoped, because there was no point in wandering off into the darkness.

I sat back down and watched. For hours we watched the fire get closer and revised estimates of speed and arrival time. They were certainly all wildly inaccurate, but there was no doubt that as the flames became more clearly defined, the smell of burning forest became stronger and pieces of burned stuff drifted over our heads. We watched, guessed and felt frighteningly helpless. There was no point in discussing options as there was only one. If the fire came around the lake, we would wait until the last possible minute, wade into the water and hope for the best. It was not much of a plan, but it would save our lives, we hoped.

Around midnight, with the flames very close to our lake, which appeared much smaller now than when we had arrived, the wind dropped. Then it picked up again, but now against my left cheek, not full face. The fire began to move away to the east. We jumped up and danced on the end of the dock, and very nearly ended up in the lake anyway. Eventually, we tried to sleep, without much success.

The float plane picked us up the next day and flew us to Fort Chipewyan where we watched the water bombers unsuccessfully attack our fire. At dusk, a larger plane picked us up to fly to Fort McMurray. We told the pilot the story of the fire and he offered to fly in close and take a look at it. From the air it was vast and the hot air currents tossed our plane about. There was no sign of our cozy cabin by the lake.

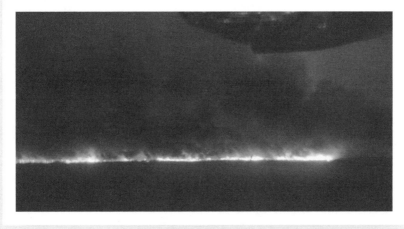

Photo by John Wilson

*Looking back from the plane with relief at the fire that caused so many anxious moments and strained one of my predictive capacities.*

# PART 2

## BIRTH AND CHILDHOOD: FOUR TO TWO AND A HALF BILLION YEARS AGO

## CHAPTER 3

# HAS IT ALWAYS BEEN THIS WAY?

What seest thou else
In the dark backward and abysm of time?

—Shakespeare, *The Tempest*

God said, "Let the waters under heaven come together into a single
mass, and let dry land appear." And so it was. God called the dry land
"earth" and the mass of water "seas," and God saw that it was good.

—Genesis 1:9–10
New Jerusalem Bible

On THE BANKS OF THE MEANDERING ACASTA RIVER, about 350
kilometres north of Yellowknife, there sits a hut. A sign proclaims, "Acasta
City Hall," but there is no Acasta City, there are just black flies and caribou
and an endless glacially scoured, rolling landscape of reindeer moss,
stunted black spruce trees, and irregular lakes. The occasional visitors to
this place leave their tools—hammers, boats and assorted camping
gear—in the hut rather than lug them in and out each trip. The equipment's quite safe, as there's no one around to steal it. The reason for even
these few visitors to this remote place is an outcrop of strangely banded

rock on a small island in a nearby lake. Neither the lake nor the island is impressive enough to warrant a name, but it is here, where perhaps on the third day of Creation the first dry land appeared out of the water, that the abysm of time is at its darkest and most backward.

The Babylonian creation myth, Enuma Elish, is known from seven clay tablets discovered in the library of Ashurbanipal beneath the ruins of Nineveh. It was written down some twelve hundred years before Christ, but certainly is much older. The tablets tell the thrilling story of Tiamat, the goddess of watery chaos, who had a penchant for creating invincible weapons. She also, according to the tablets

> "... spawned monster-serpents,
> Sharp of tooth, and merciless of fang;
> With poison, instead of blood, she filled their bodies.
> Fierce monster-vipers she clothed with terror."

Tiamat rampaged on until she came up against Marduk, the god of wind and storm. They battled furiously until Marduk

> "... seized the spear and burst her belly,
> He severed her inward parts, he pierced her heart.
> He overcame her and cut off her life;
> He cast down her body and stood upon it."[7]

The beauty of creation myths is that they are open to wide metaphorical interpretation. They are also, like the tale of Marduk and Tiamat, universally violent. The world was born out of chaos and disorder and it is tempting, when you are standing on the small island by Acasta City Hall, to see evidence of this chaos in the rock at your feet.

The rock is called the Acasta Gneiss and it looks as though it has been through the titanic struggle between Marduk and Tiamat. Bands of pale minerals swirl through it, thickening and thinning, bending and folding back on themselves as if someone has stirred a wooden spoon

*Geologists Bill Padgham and Sam Bowring (left to right)*
*overlooking rocks from the beginning of time – the Acasta Gneiss.*

through a vat of melted light and dark chocolate. Yet this is solid rock and it took immense amounts of heat, pressure, and time to make it act like liquid chocolate.

Gneiss is a high-grade metamorphic rock, which means that it formed under conditions of extreme heat and/or pressure. It generally consists of bands and lenses of alternating light-coloured granular and darker flaky minerals, hence the melted chocolate look. Its overall composition is close to that of granite, the main component of Earth's crust, and it can form through the cooking of both sedimentary and igneous rocks.

Gneiss is a common rock and its distinctive swirls can be seen on every continent. What is important about the gneiss on this unassuming northern island is that it is the oldest gneiss in the world. In fact, at 4.03 billion years old, the Acasta Gneiss is the oldest rock so far discovered on the face of the planet.

---

### Rock and Minerals: The Foundation of Geological Studies

Geologists find in rocks clues to where they came from, the movements of blocks and plates, and the conditions deep below the surface. Rocks are made up of minerals, and a rock's properties are determined by the minerals it contains. A mineral is a naturally occurring, inorganic solid crystalline material with a specific chemical composition. Minerals are made up of elements (like diamond, which consists entirely of carbon) or combinations of elements (like quartz, which is made of silicon and oxygen, or pyrite [fool's gold], which is made of iron and sulphur).

There are three types of rock: igneous, sedimentary, and metamorphic. Igneous or "fire rocks" form by the cooling of molten magma. If they cooled and crystallized quickly at the surface, they are fine-grained, extrusive igneous rocks; if they cooled and crystallized slowly below the surface, they are coarse-grained intrusive igneous rocks. Dark-coloured igneous rocks such as basalt are rich in dark-coloured minerals; they are termed "mafic." Light-coloured igneous rocks such as granite are rich in light-coloured minerals; they are called "felsic." Igneous rocks can form in various ways, such as lava flows or volcanic outpourings along mid-ocean ridges (extrusive and mafic); or as large bodies in the crust ("plutons," intrusive and felsic) that can be brought to the surface by tectonic activity or erosion.

Sedimentary or "water rocks" are the products of weathering and erosion, and originate as sediments. Some, such as limestone, are the remains of reefs formed by organisms that live in the water and make shells from calcium carbonate. They are usually found in layers that formed in shallow or deep seas or oceans. Often, sedimentary rocks are subsequently bent, stretched, and twisted into exquisite patterns. Sedimentary rocks can provide many clues to climate conditions in the geological past.

Metamorphic rocks or "altered" rocks are generated by the effects of high pressures and/or temperatures at considerable depths below the surface on pre-existing rocks of any kind. The degree of metamorphism, low- or high-grade, varies with the pressure and temperature of formation. Slate is an example of a low-pressure metamorphic rock. Gneiss, such as the world's oldest rock, the Acasta Gneiss, is an example of a high-grade metamorphic rock.

A map based on rocks at the surface, such as a normal geological map, provides a lot of information simply on the basis of the rock types and the minerals contained within them. But much more can be determined by further analysis. Chemical analyses provide a means of identifying volcanic rocks that were formed in different ways, or can be used to indicate the type of rock generated by varying geological processes. Different minerals within a rock can be used to establish the pressures and temperatures at which the rock crystallized, whether there was any subsequent melting and recrystallization, and at what depth a metamorphic rock formed. Even the process that brought that rock to the surface can be determined.

---

Humankind seems to have an abiding passion for working out how long its home has been around. Way back in the sixth century, St. Gregory the Great, witnessing the "terror with which men await the end of the world" decided to "chronicle the years already passed, that thus one may know exactly how many have elapsed since the earth began."[8] Gregory was wildly wrong, but so was everyone else, and despite the difficult nature of the undertaking there was no shortage of estimates. The first step toward a realistic guess was to realize that a long time was needed for Earth to form.

Prior to the nineteenth century, our ancestors lived in a world where almost anything was possible. Catastrophes, such as Noah's flood, had formed the world at the behest of an all-powerful, all-seeing, super-natural Creator. The nascent science of geology took on these notions and looked for instant, catastrophic causes for everything from moun-tains to gorges. But two Scotsmen were coming up with a different idea.

James Hutton in the 1790s, and his major fan Charles Lyell in the 1840s, saw things differently than the catastrophists did. Given enough time, they reasoned, the normal processes that one could witness every day, such as erosion, deposition, and volcanic eruption, could account for the landscape all around—mountains rose slowly from oceans, only to be worn down, grain of sand by grain of sand. Cataclysms became unnecessary.

The idea that the present was the key to the past became known as "uniformitarianism." Not only did it define geological thinking for a century and a half, but the idea as it applied to biology got a young Charles Darwin thinking about how different species of birds and ani-mals might arise—Darwinian evolution is little more than biological uniformitarianism.

Uniformitarianism is a comforting idea. While the Babylonian poets wrestled with wrathful gods destroying what they didn't like in cata-clysms of fire and flood, modern poets of romantic inclination could lounge about on hillsides contemplating the nature of time and human destiny in the gently changing landscape. It is much less threatening to watch grains of sand wash off a hillside and begin their long journey to an oceanic sedimentary basin than it is to contemplate an all-powerful deity who works in mysterious ways. Of course, even in a uniformitarian world there are disasters—volcanic eruptions, tsunamis, earthquakes—but they are local phenomena, minor blips in an orderly Earth. You do need to be a very optimistic uniformitarian to feel that the meteorite impact that destroyed the dinosaurs was a minor blip, but, overall, things seem to happen gradually over very long periods of time.

Uniformitarianism took hold, and with it the thought that an almost inconceivably long time was needed for all those grains of sand to find a way to the ocean. Lyell and his colleagues were not the first to postulate a great age for Earth—the mythic Hindu Vedic cycles of creation and destruction each lasted 8.64 billion years—but popular opinion in Lyell's day leaned toward everything beginning on an October day slightly more than six thousand years ago.

The October-day idea was the work of a seventeenth-century Irish bishop, James Ussher, who spent twenty years of his life meticulously delving through the Bible and over ten thousand other volumes to calculate it. The amount of work Ussher put into his guess convinced a lot of people that he must have got it right, but there was also a strong disincentive to come up with a different guess.

A hundred years before Bishop Ussher published his results in 1650, Bernard Palissy, a French writer, surveyor, garden designer, religious reformer, and potter, with a penchant for taking casts of dead lizards and snakes and sticking them to plates, pre-empted Hutton by observing that rain, wind, and tides appeared to have formed much of the landscape around him and, since they worked very slowly, this indicated a vastly ancient Earth. For this and other heresies, Palissy was condemned to death at the age of eighty in 1588 and died of starvation and maltreatment in a dungeon of the Bastille two years later.

The scientific community took some time to get over Palissy's fate, but many scientists began thinking in terms of tens of millions, rather than thousands, of years. They employed many methods in an effort to work out the planet's age, its cooling rate (24 to 40 million years), sedimentation rates (1.6 to 3 million years) and the salt concentration in the ocean (90 to 100 million years).

The most widely believed guess was by Lord William Kelvin, a Scottish mathematician and physicist with a very high opinion of himself and a disconcerting habit of making bald statements that were soon discovered to be wildly off. For example, in 1895 he stated categorically that "heavier-than-air flying machines are impossible."[9]

Despite the vastly greater lengths of time obviously required by Lyell, the uniformitarians, and Darwin for things to happen, Kelvin, working

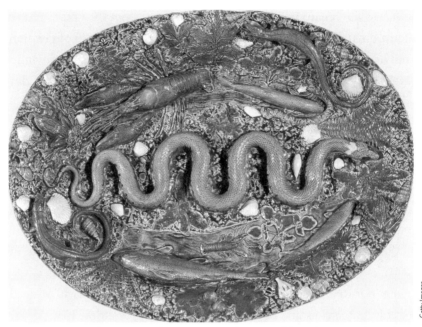

*Assorted animals sacrificed in the name of art. One of the unfortunate Palissy's plates.*

Getty Images

from the assumption that Earth had cooled from a molten ball, calculated its age at one hundred million years. Locked into his cooling scenario, Kelvin held dogmatically to his date against all comers. It was a reasonable guess, if you didn't have a source of new heat to prolong the process. Cruelly, Lord Kelvin lived long enough to see the discovery of radioactivity (capable of providing a heat source he could not have imagined), and the Wright Brothers lift off at Kitty Hawk.

Radioactivity was discovered by Henri Becquerel in 1896 and it didn't take long for someone to see its potential for finding the dates and ages of rocks and artifacts. In 1907, the year Kelvin died, the alliterative Bertram Borden Boltwood calculated an age of 2.2 billion years for Earth. He was still a long way off, but at least western scientists were now talking about the right order of magnitude—finally, the Europeans had caught up with the ancient Hindu Vedas!

Even with good techniques, the problem with dating the formation of Earth is that there is precious little left from that time to date—

intense early meteorite bombardment and billions of years of plate tectonics have destroyed whatever there might have been. There is more old rock on the Moon because the Moon's surface isn't geologically active, and dates of rocks from the Moon go back to 4.44 billion years.

The best guess we have for the beginning of things in our solar system is 4.566 billion years. This comes from dating tiny rock inclusions in the Allende Meteorite that plummeted to Earth in Mexico, early one morning in February 1969.

So Boltwood got about halfway there, and the search for the oldest rock on Earth has continued unabated. The Acasta Gneiss is the current titleholder at 4.03 billion years, but the crown is slipping. Zircon crystals from a much younger (a mere 3.3 to 3.7 billion years old) sedimentary rock in Australia have been dated at 4.4 billion years, and they must have come from somewhere, so there is, or was, a crustal rock present on Earth around 4.4 billion years ago, a date that takes us very close to the oldest dates of rocks from the Moon. Recent research on rocks from the shores of Hudson Bay suggests a possible age of 4.28 billion years, which, if confirmed, would topple the venerable Acasta Gneiss from its pedestal.

To the average person, this must all seem a bit like alchemy. Do "87Sr/86Sr ratios of 0.712" and "cumulative frequency REE diagrams" mean any more to us than "Ouroboros" and "alkahest" did to a medieval peasant? Possibly not, but at least we can find out what this all means. The alchemists treasured secrecy but modern scientists often struggle to make their work understandable to the layperson. Of course, there's a fine line here, because a scientist must balance between opaque complexity and erroneous simplicity. As Einstein put it, "Things should be made as simple as possible, but not any simpler."

So, how is this alchemy of time, this geochronological dating game, performed? In principle, it's very simple. Radioactive minerals change with time: they lose pieces of themselves in the form of radiation and so become something else.

As a child, I used to take a piece of toffee, cut it in half and eat one half. I would then take the remaining half and do the same again. By continuing the process, I reasoned, I would have increasingly small pieces each time, but I would never run out of toffee. There were a number of flaws in my rationale, not least being that I would lose patience and eat the last piece of toffee long before I tested my everlasting toffee theorem. Had I only known about radioactive decay.

### The Dating Game

The rock record, incomplete though it may be, provides geoscientists with the means of determining the fourth dimension of geological studies—time—without which they could not understand the development of our planet. Here, sedimentary rocks play an important role. Generally speaking, sediments are deposited as horizontal layers; variations from this typically indicate that the rocks were deformed by some tectonic process. In the absence of any type of deformation, a given layer of sediment should be younger than the layer below it and older than the one above it. This provides a form of relative timing. Sediments from the Phanerozoic eon contain fossils. Paleontology is the study of life in past geologic time based on fossils of animals and plants. Early on in the study of geology, paleontologists and other geologists found that fossils, most of which are unique to a particular geological time period, could be used to establish relative timing and to correlate rocks of different ages at different locations on the globe. William Smith's 1815 map, "A delineation of the Strata of England and Wales with part of Scotland" (see page 27), employed the first use of these concepts. Since then, paleontology has been applied and refined to generate a complete geological time scale based on the fossil record (now known to go back about 600 million years).

But the geological time scale derived in this way only represented relative time. The actual time for a geological process, such as laying down a sedimentary layer, or how long a particular fossilized species had existed, could not be determined. The discovery of radioactivity in uranium in 1896 by the French physicist Henri Becquerel, and the subsequent discovery of the radioactive element, radium, in 1897 by the French chemist, Marie Curie, changed all that. And a Canadian scientist led the charge. Ernest Rutherford, working at McGill University in Montreal, recognized that radioactivity could be used to measure the age of a rock. From measurements in his laboratory, he measured the age in years of a uranium mineral. Since then, the technique has been refined to the extent that rocks three billion years old can be dated to a precision within one to two million years.

Zircon is the mineral most commonly dated due to its resistance to modification by temperature, tectonic activity, or chemical processes. However, other minerals are also commonly used and provide equivalent or complementary dating information. Apatite, a calcium phosphate mineral, occurs as an accessory mineral in igneous rocks, metamorphic rocks, and ore deposits. Since apatite includes some uranium atoms, it also can be used for dating by the uranium-lead method. The mica minerals are potassium aluminum silicates with a distinctive one-directional cleavage such that they break into thin, elastic flakes. The common mica forms found in rocks are biotite, a dark-coloured iron/magnesium-bearing mineral, making it of the mafic family; and muscovite, a white or colourless mineral without the iron/magnesium content, making it of the felsic family. Both biotite and muscovite contain potassium and rubidium atoms and thus can be used for dating by the potassium-argon method and the rubidium-strontium method.

These and other minerals are stable only under particular conditions of pressure and temperature. Thus, earth scientists can date geological processes such as the cooling of a particular rock below a specific temperature or the time at which a specific rock reached a certain depth (and corresponding pressure) below the surface.

Let's take a step back to understand more about radioactive dating. Minerals are made up of chemical combinations of elements with a specific internal atomic structure. Elements are made up of atoms, the smallest unit of an element that can exist and still show the characteristics of that element. A simple view of an atom is that it includes a densely packed nucleus of protons and neutrons surrounded by a cloud of electrons. The number of protons in a given element is constant, and thus defines the element. However, the number of neutrons in an atom can vary. Thus, the atomic mass of an element, which is the combined weight of the number of protons and neutrons, can vary depending upon the number of neutrons. For example, lead-204, the common form of lead, has eighty-two protons but two less neutrons than lead-206 and three less than lead-207, both of which form as the result of radioactive decay. The three types of lead are examples of isotopes; they all have the properties of lead and the same number of protons, but differ from each other in their atomic mass.

While the nuclei of most elements are stable, some elements have nuclei that are naturally unstable and disintegrate through radioactive decay. These can be used for dating purposes. Carbon-14, uranium-235 and uranium-238 are examples. When the number of electrons (negative charge) equals the number of protons (positive charge) in an atom, the atom is electrically neutral. However, most elements can lose or gain electrons in the outer parts of their cloud, resulting in an atom with an electrical charge, called an ion. The ionic charge and the ionic size, indicative of the number of neutrons in the nucleus and the number of electrons in the surrounding cloud, provide the controls on how elements combine together to make solid minerals.

To date a rock, minerals such as zircon, apatite, muscovite and/or biotite must be separated from the rock sample. This usually involves crushing the

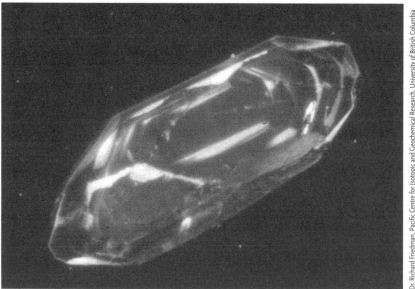

*A zircon crystal, hoarding the secrets of time. This zircon is about 0.3 mm long, much smaller than the size of a pin head.*

rock and then manually searching for the minerals that can be dated. Once the minerals have been separated out, they are prepared by physical and chemical means for insertion into a mass spectrometer. For example, they may be put into solution and deposited on a filament, then evaporated by heat; or they may be blasted by a laser to generate a gas. A mass spectrometer detects ions of different masses so the relative abundances of different isotopes can be measured, from which the mineral's age can be determined. Since there are errors associated with the measurements and calculations, many mineral samples from a given rock are usually dated to provide statistical validity to the results.

All radioactive elements have a half-life, the time in which half of the original material will decay to something else. Some exotic elements have half-lives on the time scale of my toffee-cutting experiments, but others decay on a geological time scale. For example, if you have one gram of radioactive uranium-235 (U-235), after 704 million years, you will be left with half a gram of U-235 and half a gram of lead-207 (Pb-207). After 1.408 billion years, you will have one-quarter of a gram of U-235 and three-quarters of a gram of Pb-207, and so on.

This is the opposite of the alchemist's dream of turning lead into something useful and makes stockpiling radioactive minerals a poor long-term investment, but it is a useful tool for looking at the past. Since the radioactive decay of U-235 to Pb-207 merrily goes its own way regardless of heat, pressure and the petty world around it, it is a perfect measure of passing time. The proportion of U-235 to Pb-207 will give you a measure of the time that has passed since the original moment, when all the Pb-207 was U-235.

Of course there's a catch. You have to assume that there was only U-235 to begin with, and that neither element has been either added or subtracted over the millions of years you are examining, but given the stability of radioactive decay and the elements involved, this is usually a reasonable assumption. And there are checks. Uranium-238 (U-238) decays with a half-life of 4.468 billion years to lead-206 (Pb-206); potassium-40 (K-40) produces argon-40 (Ar-40) with a half-life of 1.251 billion years; and rubidium-87 (Rb-87) ends up as strontium-87 (Sr-87) after a staggering half-life of 48.8 billion years. Each system has flaws. For example, argon is a gas and is lost from minerals when they are heated, so that the K-40 to Ar-40 dating system is only a measure of the time since a mineral was last heated. Nevertheless, if two separate processes give the same date, we can have a significant level of confidence that we are on the right track, especially if we are using zircon.

Zircon has a long history. In Hindu lore, the ultimate gift to the gods was a glowing tree covered in gemstones and zircon leaves; in the Middle Ages owning zircons made you wise and prosperous and helped you sleep; today it's your birthstone if you were born in December. The name is derived from two Arabic words, *zar*, meaning gold and *gun*, meaning coloured, though zircons are not often gold-coloured. Variously coloured zircons have gem value and clear ones can be difficult, even for an expert, to distinguish on sight from diamonds. This means that they have been used over the years for both innocent and villainous purposes. But zircons have characteristics that make them extremely valuable to geologists.

Zircon ($ZrSiO_4$) occurs in small quantities in almost all igneous rocks and usually has uranium as an impurity replacing a few zirconium (Zr) atoms. Zircons also tend not to incorporate lead atoms, are resistant to high temperatures, are hard, and are chemically inert. It's tough to mess with zircon and this makes it almost ideal as a mineral for dating. All you need do is take a zircon crystal, discover the ratios of U-235 to Pb-207 and U-238 to Pb-206, you have two pretty accurate measures of the age of the mineral and hence the time of cooling of the igneous rock that contains the zircon.

The only difficulty with zircon is that it tends to cover itself with overgrowths, giving the crystals a zoned appearance. An early problem in dating whole zircon crystals was that any date was an amalgam of the original core zircon and the later overgrowths. More recent techniques can now work with minute pieces of individual grains, thus obtaining dates for the core grain as well as the overgrowths. The former give the age of the rock and the dates of the overgrowths indicate the dates of other active geological processes.

It's all very clever, but is today's alchemy any more significant than the ancient? Does it matter if we know within a few million years how old our planet is, or if we can stand with assurance on the oldest surviving rock? On a day-to-day basis, no, but in terms of understanding our world, yes.

Arthur Holmes, arguably the greatest geologist of the twentieth century, developed a splendid analogy for the human study of geological time, perhaps after reading through Einstein's theory of relativity. What if, he postulated, we all lived on the face of an old-fashioned alarm clock and our life span was but one second? Once we began, as a society, to be aware of our world, we would notice that the second hand of the clock was moving. After several generations of carefully documented study, we might feel able to say with some certainty that the minute hand was moving. Perhaps some brilliant scientist on our world would postulate that the hour hand was moving and even how the three hands might relate to one another. Perhaps that is the stage we are at with the theories of plate tectonics and supercontinents—glimmerings of an understanding of the clock hands, but no real idea of how the springs, gears, and wheels work and interlock behind the face. And will we ever begin to guess what the alarm mechanism is for?

The Acasta Gneiss and the events that we are beginning to discern as having occurred billions of years ago are, perhaps, the equivalents of the movements of the hour hand on Holmes' clock face. Maybe, not too long before the Acasta Gneiss formed, as the lives of planets are measured, was when the hour hand began to move.

Once upon a time there was a solar nebula, a swirling cloud of hydrogen and helium. As it spun, a dense centre, a primitive protosun, began to form. As this happened, an intense solar wind swept out from the centre, sucking all gas and volatiles away from the inner solar system, leaving behind the heavier elements. Particles of these elements eddied around, bumping into each other. When the speed and the angle of impact were right, the particles stuck together and clumps formed. These clumps grew. The larger they became, the greater their gravitational attraction, and so they grew ever faster, eventually becoming planetesimals, or small planets. These continued to sweep up dust and stick together until, around four and a half billion years ago, most of the debris had coalesced to form a few large round lumps, one of which was destined to become Earth. But it was not an easy infancy. Almost as soon as cooling began, another lump, about one-fifth the size of Earth, collided with us, melting what had solidified and throwing off a sizable chunk. The chunk was captured by the stronger gravitational attraction of Earth and became our Moon.

Things settled down a bit after that. There were still many impacts, but the cooling continued, water collected, a thin atmosphere formed, and, according to some scientists, life began its strange and mysterious journey. Unfortunately, between about 4 and 3.85 billion years ago, something upset the fragile balance of the developing system and sent swarms of asteroids plummeting towards the Sun. This shotgun bombardment is what gave the Moon its familiar cratered face. On the infant Earth, the larger impacts would have blasted off the early atmosphere, vaporized the water, and fried any life forms that might have existed.

In the relative quiet that followed, an atmosphere of carbon dioxide and methane formed, surface temperatures reached around 70°C, mountains poked out of the planetary ocean and continents began to form and be destroyed. The processes we recognize today as plate tectonics became established, but exactly when? It is against this background that Lithoprobe looked at the Slave Province and, to the east, the Superior Province.

Plate tectonics is both inevitable on a geologically active planet, and essential for life to develop. Without the dance of the continents and the continuous creation and destruction of crust, a world dies, but is there only one way for plate tectonics to occur? This is one of the fundamental questions that Lithoprobe asked.

Oceanic crust forms along mid-oceanic ridges, but continental crust forms where the oceans are destroyed. Subduction of oceanic and continental crust beneath other pieces of crust is essential to plate tectonics. It might be nice for the residents of Vancouver Island to imagine a world without subduction, because then they wouldn't have to worry about "the big one" disturbing their idyll in the near future, but without subduction, where would the new plate formed at the mid-oceanic ridges go? Without subduction, plate tectonics as we have come to understand it couldn't operate. Without plate tectonics continents couldn't form into stable platforms with nutrient-rich continental shelves where life could develop and blossom before crawling out onto the land. Without plate tectonics, we wouldn't be here to wonder about our Earth.

Subduction depends on a fine balance of temperatures. The cool, less dense crust floats on a warmer, denser mantle. If the temperature differential increases, say by an increase in the temperature of the mantle, the crust becomes more buoyant and, therefore, more difficult to force down into the mantle. Subduction becomes impossible.

Most theories of what the world was like when the Acasta Gneiss was forming postulate a warmer mantle, warm enough to prevent subduction. There was a crust, but it was basalt, like the crust beneath the oceans today or on the Moon's *mare*. There were no continents because continents are made mostly of granitic rocks, which form in two primary ways: from the metamorphism of sediments from continental erosion (which didn't yet exist to erode), or from the metamorphism of basaltic rocks in the presence of water when they are dragged down into a subduction zone—no sediments, no subduction, so no granitic crust. Simply, plate tectonics only began to operate when there was ample water and a cooler mantle, which allowed granitic rocks to accumulate into something stable and big enough to be considered a continent. Just how far back can we definitely push plate tectonics?

Plate tectonics operating on the Acasta Gneiss, four billion years ago, would require a major rethink in the postulated physics of planetary formation. It would be, quite literally, Earth-shaking. Hard evidence is needed. Unfortunately, that is difficult to come by in an area like the

Slave Province, which has been subject to such a long history of change. But there are clues.

A basic element of the theory of plate tectonics is the presence of continents. Without them, nothing else is possible. So the first step in pushing plate tectonics back into the nearly fathomless abysm of time is to find out if there were even continents four billion years ago.

Continents are granitic. Gneiss is granitic. Including the Acasta Gneiss, there are about thirty-five fragments of very ancient crustal rock in the global record that far back. There may be more, as witnessed by the 4.4-billion-year-old relic zircons, but are these fragments of true continents? Unfortunately, the jury is still out. Certainly the conditions for granitic crust to form—high pressure, high temperature, and large quantities of liquid water—existed far back in Earth's past, but how large were those fragments?

There are tantalizing suggestions. All the small pieces of very ancient crust scattered across the Slave Province fit a pattern. Wherever the top of the ancient gneiss is seen, it is weathered and uneven. That means that it was exposed to the atmosphere for a long time many millions of years ago. On top of the gneiss lie sediments and ancient lava flows that date to 2.8 billion years ago.

This similar history suggests a common heritage of erosion, burial in sediments, and then covering with lava. That is the sort of thing that happens when a continent breaks apart, but what was happening to the Acasta Gneiss and its neighbours for the billion or so years prior to that?

More clues. At the eastern edge of the Slave Province the rocks are missing, just as if they have rifted off and drifted away. This is common to all the ancient pieces of crust. Does that mean that they were once joined together? As Wouter Bleeker of the Geological Survey of Canada put it, "If you see the head of an elephant cut off at the neck, you know there must be a body lying around somewhere."[10] But where is the body?

Are we looking at remnants of the very first supercontinent? Some Earth scientists think so, and have even named this pre-three-billion-year-old supercontinent, Ur. The likelihood of Ur's existence is tentative, to say the least. More likely, we are looking at several smaller micro-continents with distinct histories. This is the theory that Bleeker favours and he has gone as far as to name three of them: Sclavia, Superia, and Vaalbara. Sclavia is the Slave Province, along with similar rocks in India, Wyoming, and Zimbabwe. Superia is based on early rocks from the Superior Province and Vaalbara is built around the Kaapval Craton in South Africa.

### The Slave Province

The Slave Geological Province is a relatively small chunk (about seven hundred by five hundred kilometres) of Archean ancient crust situated in remote north-western Canada. Although the area is remote, many geological studies, which have provided considerable details on the current structure and past development of the craton, have been carried out. The map (page 104) shows the four main types of rocks that form the Slave Province. Gneissic rocks (dark grey) represent the old foundation of the craton, including the Acasta Gneiss, upon which the younger sequences were built in ways that are not yet fully understood. Rocks from this old foundation represent billions of years of crustal development. The Acasta gneisses formed 4.03 billion years ago, and the youngest gneisses are about 2.85 billion years old.

For the next one hundred million years or so, this earlier foundation of rocks was probably flooded by shallow seas, and during this period some characteristic rocks were deposited on the sea floors. Then, about 2.73 billion years ago, there was a vast outpouring of lava-like basalt. This lasted for about thirty million years and formed layers of volcanic rocks up to six kilometres thick. The medium-grey shaded areas on the map represent the metamorphosed remains of these volcanic rocks now exposed at the surface, the second main rock type. Toward the end of the period of massive volcanism, rocks deeper in the crust started to partially melt. The molten rock, lighter and thus more buoyant than the rest of the crust, worked its way upward to form large bodies of granite rock called plutons. The earliest of these intermingled with the volcanic rocks. However, plutonic activity continued for about another one hundred million years until about 2.6 billion years ago. Thus, the third main type of rock forming

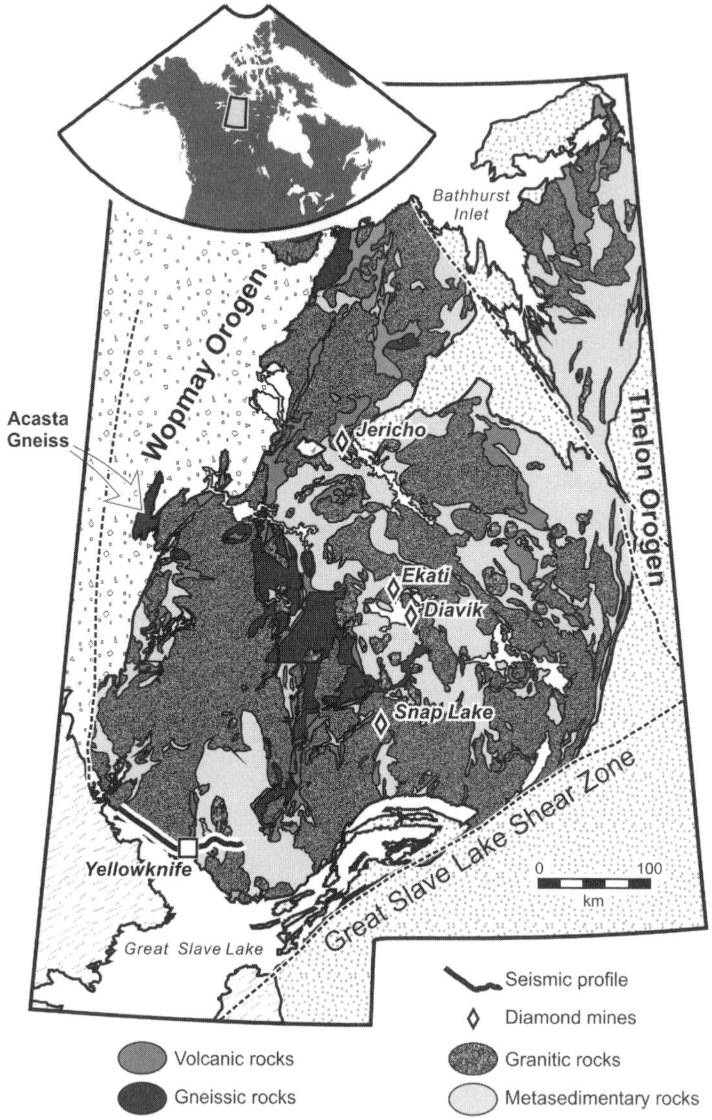

A simplified geological map of the Slave Province and surrounding regions. The legend identifies the different rock types. The Archean cratonic core is bounded by fault zones (dashed lines), younger Proterozoic mountain-building episodes called orogenic belts (e.g., Wopmay orogen to the west; Thelon orogen to the east), and Phanerozoic cover of the Western Canada Sedimentary Basin to the south and west of Great Slave Lake. Diamonds show the locations of the Ekati, Diavik, and Snap Lake diamond mines, which are currently in production, and the Jericho mine, currently in development. The seismic reflection line, along which the data of the next illustration were acquired, is shown passing through Yellowknife. The inset shows the location of the main map on an outline of northern North America.

the Slave Province is granite and related rocks, shown as the textured grey pattern on the map.

This ever-changing landscape probably had a lot of topographic variation as well. Erosion wore down some of the existing rocks, forming sedimentary rocks that are now found exposed throughout the Slave Province. These were subsequently metamorphosed by being buried and then brought to the surface again by erosion and buoyant forces. These metamorphosed sediments are the fourth major rock type in the Slave, shown on the map by the light grey shading.

The detailed processes and rock types that established the Slave Province were far more complex than this simple discussion indicates. The full story is still not known even to the scientists who work there. As far as they can tell, the Slave, as we now see it, is only a small fragment of a presumably much larger craton, which was broken up by rifting, forming other fragments that have since drifted far and wide. For example, one theory suggests that the Wyoming craton of the northwestern United States, the Dharwar craton of India, the

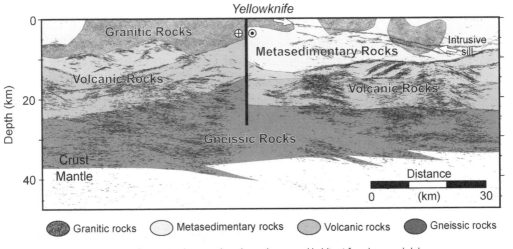

Computer-generated seismic reflection record section (short, thin, and numerous black lines) from data recorded along the seismic profile shown on the map of the Slave Province. A simplified interpretation is overlaid on the seismic data. The shading corresponds to the four major rock types indicated on the map. Based on geology and dating, the volcanic and granite layers have been separated into two slightly different types. For the volcanic rocks, the two types are shown as slightly different shades of medium grey. The thick line below Yellowknife indicates a transcurrent fault; the circle with a dot indicates the block on the right moved toward the reader; the circle with a plus sign indicates the block on the left moved away from the reader. The nearly continuous black line in the upper right (identified by arrows) is interpreted to represent a mafic sill that intruded into the upper crust in Proterozoic time, long after the Slave Province was established and quiescent, because such a sill is exposed at the surface.

Zimbabwe craton in Africa and the Slave may at one time have formed an Archean supercontinent. Now the Slave Province is bounded by fault zones and younger blocks with which it collided after two billion years ago.

This story of the Slave Province is based on studies of rocks exposed at the surface. However, to provide a lithospheric view to great depths, geophysical techniques are required. The only all-season access road winds along the north shore of Great Slave Lake and ends about seventy kilometres east of Yellowknife. For cost-effective seismic reflection and refraction studies, such roads are necessary. Thus Lithoprobe seismic studies have been confined to the south-western part of the Slave Province. The seismic reflection image, combined with the known surface geology, can be interpreted in terms of the four major rock types, but shows details of probable deformation, stratigraphic layering and/or other features within these major units.

---

But whether these incredibly ancient continental fragments acted in any recognizable way is a very different question. Was there subduction in Acasta times? Were there spreading ridges? Did primordial mountain chains rise and fall as protocontinents crashed together? Did those continents grow by horizontal accretion as other plates were destroyed in subduction zones along their edges? It's all speculation. What would be nice would be to find a fossil subduction zone.

Modern subduction zones angle down relatively sharply beneath continental edges. If one were to be found in the pictures that Lithoprobe took of the rocks beneath the Slave Province, it would show up as a dipping boundary between two different crustal pieces.

Unfortunately, no one has so far found a four-billion-year-old subduction zone. In fact the evidence from beneath the Slave Province suggests something different—the boundaries that might represent the divides between crustal plates coming together are almost horizontal. This doesn't fit with subduction, but rather a rafting of one thin continent beneath another. The continents grew, but vertically rather than horizontally.

So questions still hang in the air around the existence of plate tectonics back when Marduk and Tiamat were having at each other. Certainly there were floating fragments of granitic crust four billion years ago. Probably they moved about, joined together and broke apart, but not in the same way as continents do today—the protocontinents were too thin and the mantle below was too hot to allow the processes we recognize to occur. So if not then, when? When did recognizable plate tectonic action begin?

Most of the remaining really ancient crust on the planet is something called granite-greenstone terrane. It comprises broad areas of generally granitic rocks separated by linear belts of greenstone rocks. The greenstones are so called because they characteristically contain a collection of green metamorphic minerals, the very ones that so confused me on my first day on the job in Zimbabwe.

Understanding these rocks is important since they host more than half the gold in the world, large amounts of other metals, and a healthy proportion of diamondiferous kimberlite pipes. Granite-greenstone terranes can be studied in many places across the world, but the best place is in Canada. Lithoprobe scientists couldn't pass up a chance like this.

The Superior Province, which stretches between Hudson Bay and Lake Superior, is the largest and best-exposed piece of ancient crust around. Its formation dates to around 2.7 to 3 billion years ago and it is the core onto which the other pieces of the North American jigsaw accreted.

Traditional theory explained granite-greenstone terrane evolution through mostly vertical movements. The wide bodies of granitic rock welled up, squishing sequences of sediments and volcanic rocks into linear belts that looked, in profile, rather like the deep keels of boats. Heating during this process produced all the green minerals. Plate tectonics is characterized primarily by horizontal, rather than vertical, movement of pieces of crust. If Lithoprobe studies could confirm that horizontal movement of the pieces that came to be the Superior Province was a major factor, this would strongly support the role of plate tectonics in its formation.

## The Superior Province

The Superior Province forms the central Archean core of North America. It is the largest exposed craton in the world and it has generated enormous mineral wealth for Canada. As a result, the Superior Province is well studied, particularly in mining regions, but continuing studies are needed, and are being done. During the late 1980s and into the early 2000s, Lithoprobe and the National Mapping Project of the Geological Survey of Canada combined to compile further geological data and geophysical imaging of the southern parts of the province to better understand its geological development.

The Superior Province includes rocks that are dated as old as 3.7 billion years, but most of the activity for which the rocks preserve a record occurred after 2.8 billion years ago. Since 2.6 billion years ago, the entire region has been tectonically stable. Around the margins of this region, younger blocks became amalgamated in subsequent geological time to enlarge the continent of North America to what it is today.

A first-order and somewhat unusual feature of the Superior Province is its linear system of subprovinces distinguishable by the distinctive types, ages, and structural styles of the rocks that form them. As shown by the map on page 109, these linear trends generally run east-west in the south, west-north-west in the northwest, and north-northwest in the northeastern Superior Province. Dating of rocks in the subprovinces has revealed five distinct, and very old, domains of continental crust. These fragments existed for up to hundreds of millions of years before the development of ocean basins.

The oldest remnants of continental crust are about 3.7 billion years old and occur in the Northern Superior Superterrane. To its south, the North Caribou Superterrane is a large remnant of crust that is about 3.0 billion years old and is the nucleus around which the Superior Province amalgamated. The third and fourth ancient continental fragments, the Winnipeg River and Marmion terranes, are relatively small regions that date back to 3.4 and 3.0 billion years ago, respectively. At the far southern extent of the Superior Province lies the fifth old continental fragment, the Minnesota River Valley Terrane, a domain of unknown extent but with crustal remnants as old as 3.5 billion years.

Most of these ancient continental fragments are separated on the map by distinct younger domains made of rocks that have an oceanic, rather than continental, affinity. These are dominantly the granite-greenstone terranes with rock types that resemble those of modern-day ocean floor, oceanic islands, and oceanic arcs, such as Indonesia. Examples include the Oxford-Stull Domain at

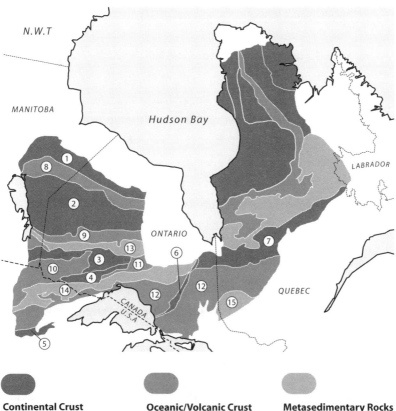

**Continental Crust**
1 - Northern Superior
Superterrane
2 - North Cariboo Terrane
3 - Winnipeg River Terrane
4 - Marmion Terrane
5 - Minnesota River Valley
Terrane
6 - Kapuskasing Zone
7 - Opatica Domain

**Oceanic/Volcanic Crust**
8 - Oxford-Stull Domain
9 - Uchi Domain
10 - Western Wabigoon
Terrane
11 - Eastern Wabigoon
Terrane
12 - Wawa-Abitibi Belt

**Metasedimentary Rocks**
13 - English River Terrane
14 - Quetico Terrane
15 - Pontiac Terrane

Map of the Superior Province showing the linear belts comprising the three main types of crustal material, indicated by the different shades of grey. Individual domains that were part of the Lithoprobe studies are identified by the numbers.

the northern margin of the North Caribou Superterrane, the western Wabigoon Terrane in the western part of the province, and the Wawa-Abitibi belt in the southern part.

Some of the continental and oceanic domains are separated by still younger features that comprised metasedimentary rocks. Examples include the English River, Quetico, and Pontiac terranes. These fifty- to one-hundred-kilometre-wide belts of metamorphosed sedimentary rocks, and granite derived from them, extend across the entire Superior Province. They formed during the mountain-building processes associated with the collision of continental and oceanic domains.

---

The entire Superior Province has been metamorphosed, but consists of broad bands, or terranes, that run east to west and are generally older in the north. Some of the terranes were originally granitic rocks, some were sedimentary rocks, and some were volcanic rocks. Lithoprobe looked at the huge data base already assembled on this well-studied piece of the ancient Earth and gathered new data on the geometry and structure underneath the various terranes to see if it fitted with a plate tectonic origin. The whole made a fascinating, forty-million-year-long story.

Some of the terranes in the Superior Province are actually older bits and pieces stuck together and are therefore called "superterranes." The oldest superterrane in the Superior Province is the most northerly, the imaginatively named Northern Superior Superterrane. To the south lies the North Caribou Superterrane, the largest continental fragment and the one that served as a nucleus for the formation of the entire province. Next come the Winnipeg River and Marmion terranes, and then the Minnesota River Valley Terrane, which contains rocks as old as 3.5 billion years.

These five major crustal fragments are separated by bands of younger, assorted oceanic crust and sediments that were seriously messed up as the continental fragments stuck together. Within each of these newer bands, distinctive minerals formed at specific temperatures, marking the time that the surrounding pieces of crust joined. The ages of these minerals give us a precise chronology for when the microcontinents came together. In detail, it is a complex story, but, broadly, here is what happened.

At 2.75 billion years ago, the five independent microcontinental fragments were separated by ocean basins of unknown size, although Winnipeg River and Marmion terranes were probably close to each other. By 2.72 billion years ago, the Northern Superior Superterrane collided with the North Caribou Superterrane. About the same time, to the south, the Wabigoon, Winnipeg River, and Marmion crustal pieces came together to form a superterrane that, twenty million years on, collided with the growing continent to the north. At 2.69 billion years ago, the Wawa-Abitibi belt joined the party and, finally, ten million years later, the Minnesota River Valley Terrane collided with the enlarged continent to complete the process.

This looks like a story of horizontal movement, of plates drifting, of continents crashing, and of mountain chains rising all across northern Manitoba, Ontario, and Quebec. Each time two pieces of crust crashed into each other, mountains formed from the squashed oceanic crust and sediments between them. Deep below these long-vanished peaks, under immense heat and pressure, the dated minerals formed, giving precise ages for the mountain building and, hence, the microcontinental collision.

To confirm the tale, the Lithoprobe seismic reflection data even managed to spot dipping structures beneath the microcontinental edges that are interpreted as fossilized subduction zones. This proves that there was recognizable plate tectonic activity around 2.7 billion years ago. Perhaps the events outlined in the formation of the Superior Province even recorded the first stages in the formation of an early supercontinent, Kenorland, much younger than the theoretical Ur, but a little more concrete in our imaginings. Kenorland would have been complete around 2.45 billion years ago and lasted for some three hundred million years before breaking up.

Whether Kenorland existed or not, plate tectonics certainly did. It may not have acted exactly as it does today, but it was probably a much more recognizable process than the one that formed the Acasta Gneiss. For example, the mantle might have been cool enough to allow subduction, but still too warm to allow the deep subsidence at continental edges that, filled with thick sequences of sedimentary rocks, are one of the characteristics of modern subduction zones. But the process of continents and oceans interacting almost three billion years ago was similar to today.

Lithoprobe and other studies have produced a lot of hard evidence that plate tectonics was up and running by about 2.7 to 3 billion years ago. Perhaps that's not the third day of Creation, but it is farther back than most people thought and well enough established that we can use it to look at the next stage in the childhood of our planet.

# African Hazards
by John Wilson

Z IMBABWE IN THE MID-1970S, although few realized it at the time, was one of the best places in the world to study greenstone belts and see the other bits and pieces that had once been attached to the Slave and Superior Craton. It was also a country in the middle of a war for independence.

The hazards of fieldwork in the African nation were many. I once had a bull kudu antelope, heavy enough to crush my Land Rover and me to pulp, leap over the hood as I merrily sped along a dirt road. Another time I walked around a granite boulder and ended up face to face with a Mozambique spitting cobra. These were minor hazards compared to the main one at that time: humans armed with AK-47 automatic rifles.

*Field work in wartime. John Wilson with field assistant in Zimbabwe, 1976.*

The war in Zimbabwe was brutal on both sides, and we young enthusiastic geologists at the Geological Survey were never allowed to forget it. We were told not to drive on dirt roads (an impossibility with the realities of field work, but I suppose it made the powers that be feel better), report to police camps before and after each day's work, carry a gun, and never drive at dusk. The last injunction was because most of the rural attacks happened at dusk to give the attackers the safety of night to escape.

Being a coward, I followed as many of the rules as I could and, more because of luck than compliance, never got into serious difficulty. The only long-term consequence of my two years in Zimbabwe is a vague unease when I sit in a lighted room at night with the curtains open—someone can see in but I can't see out. One of my colleagues was less lucky.

Jim Warren's field area was in the east of the country, too close to the particularly unsettled border area with Mozambique. One evening he was driving his Land Rover back to camp too late in the day. He came over a rise and saw a group of men by the roadside kicking a soccer ball around. Jim was suspicious. Why play soccer by the road and why were they moving so stiffly? However, there was little he could do as the road was too narrow to allow a three-point turn. Determining that the safest course was to pass the group as quickly as possible, Jim gunned the Land Rover's engine.

Gunning a Series 11 Land Rover's engine, especially if it has been subject to many years of use, doesn't do much, but partly because he was going downhill and partly because he was urging his vehicle on loudly, Jim was travelling at a fair speed as he reached the men. To his horror, he saw that the reason they had been moving so stiffly was that they were carrying AK-47s beside their bodies. These they now raised and opened fire on the speeding vehicle.

Jim hunched forward over the wheel and began rocking back and forth to try and make his vehicle move faster. Glass shattered and bullets flew through the door and into the dashboard. A bullet, or a piece of

Land Rover, tore a gouge in Jim's hand, but then he was past, bouncing crazily along the road and sweating profusely.

After several kilometres, Jim relaxed enough to be able to loosen his hands from the steering wheel. Sighing with relief, he slowed and leaned back, and back, and back. The entire back of his seat had been shot away. If Jim hadn't been leaning forward to encourage the old Land Rover to greater effort ...

## PART 3

# TUMULTUOUS TEENS: TWO BILLION TO ONE BILLION YEARS AGO

## CHAPTER 4

# GLUING IT ALL TOGETHER

Never measure the height of a mountain until you
have reached the top. Then you will see how low it was.
—Dag Hammarskjold, former Secretary-General of the UN

Gradually, very gradually, we saw the great mountainsides
and glaciers and aretes, now one fragment now another through the
floating rifts, until far higher in the sky than imagination had dared
to suggest the white summit of Everest appeared.
—George Leigh Mallory

NOVEMBER IS THE BEST MONTH to visit the plains around Mysore
in southern India. Daytime temperatures hover comfortably in the
seventies Fahrenheit. The air is clear, there is only a one in eight chance
of rain on any given day, and there is a zero chance of snow. The area
offers a fine selection of temples, palaces, wildlife preserves and the island
fortress of Srirangapatna. Unfortunately, there is nothing left to com-
memorate one of the most significant events that occurred on the plain.

Had you been near Mysore in November 1800, you would have
witnessed an extraordinary scene. A hundred-foot-long centipede-like

structure of boxes, covered by large tents and surrounded by a circus of milling workers crawled across the landscape with exquisite slowness. The whole was organized by a British officer in the red uniform coat of His Majesty's 33rd (1st Yorkshire West Riding) Regiment of Foot, who only the previous year had helped Lieutenant-Colonel Arthur Wellesley, the future Duke of Wellington, pacify the surrounding countryside by storming Seringapatam (as Srirangapatna was known then).

As your eye became accustomed to the apparent chaos, you would notice that groups of workers were engaged in specific tasks. Several were pounding stout pickets into the ground ahead of the boxes. When they were finished, twenty others would lift the centipede and move it onto the new pickets. The tent would follow and the officer would bend to carry out some arcane adjustments to whatever lay within the boxes. This operation had been repeated since October 14 and would continue until December 10 to cover 7.4321 miles—a total of 393 moves. It was brutally monotonous work, but it was vitally important, for this activity was a trial run for an extraordinary British imperial undertaking: the Great Trigonometrical Survey of India.

If you control a far-flung empire, it is a good idea to know as much about it as possible. In particular, taxation and pacification require a good working knowledge of the land, its people, and its resources. Accurate maps are important to both those who believe an empire is to be exploited and those who feel it should be civilized. The accurate mapping of Britain's Indian possessions was a daunting undertaking, but the circus-like operation outside Mysore was a good beginning step.

In the days before satellites and global positioning systems, mapping was done by surveying the ground. Two hundred years ago, this involved a primary theodolite that might weigh over one thousand pounds, several smaller optical instruments, heliotrope mirrors, lights, measuring rods, and a mountain of tents, miscellaneous equipment and supplies. In India, the gear and personnel were transported on an immense caravan of four elephants, thirty horses and forty-two camels. The surveyors and military officers rode on the elephants and horses and the

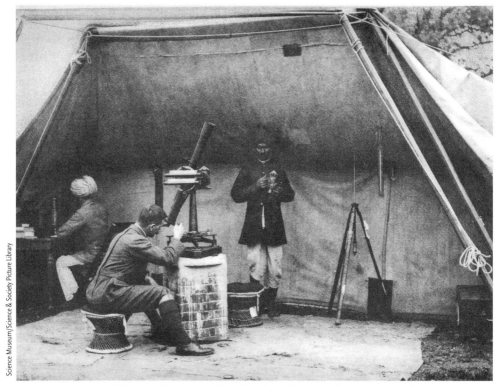

*How things were done at the height of Empire. A few pieces of equipment from the*
*Great Trigonometrical Survey of India.*

seven hundred or so labourers walked. As the caravan progressed, it surveyed lines in groups of three to create triangles, each of which had at least one side in common with a previous triangle. Most of the lines were surveyed optically but, for precision, baselines for each triangle had to be measured directly. That is what the circus outside Mysore was doing, measuring the very first baseline.

The boxes beneath the tents carried a blistered steel chain of forty links, each two and a half feet long. On either end were brass registers that could be finely adjusted by finger screws and from one end hung a twenty-eight-pound weight to keep the chain taut. At 62 degrees Fahrenheit, the chain measured exactly one hundred feet and had a correction of 0.00725 inches for every one degree of temperature variation. This magnificent, if cumbersome, measuring device had originally been presented to the Emperor of China. Not being a party to the craze for measuring things that was sweeping the western world, the Emperor had been unimpressed and rejected the gift, which eventually ended up outside Mysore.

The Great Trigonometrical Survey progressed, in fits and starts and with many sidetracks, throughout much of the nineteenth century. Its most dramatic feature was a survey line running for almost 2,400 kilometres along the seventy-eight-degree east meridian from Tinavelley at the southern tip of the subcontinent to Banog in the foothills of the Himalayas. The purpose of this line was to measure a sufficiently long curve on Earth's surface to enable an accurate determination of the size of our planet. Few now recall what that result was, but the surveyor's obsession with accuracy turned up something else.

Where possible, triangulation results on the ground were double-checked astronomically. As the line progressed north toward Peak xv in the towering Himalayas, it encompassed the town of Kalianpur and its near namesake, Kaliana, 375 miles to the north. The calculations of the distance between the towns by triangulation on the ground and astronomical means varied by 550 feet, or 5.23 seconds of arc, on the meridian.

The surveyors were not too bothered by this. A century before, the French scientist Pierre Bouguer had noticed a similar effect on survey instruments in the Andes. The theory was that the huge, dense mass of a mountain chain exerted a gravitational pull on the plumb bobs that were used to determine the vertical setting for the survey instruments, and this caused a false reading. It was not until 1855, almost ten years after the meridian line was completed, that a problem was spotted.

J. H. Pratt, a Church of England archdeacon stationed in Calcutta, made an estimate of the minimum mass of the Himalayas and thus calculated what the gravitational effects should be at Kaliana and Kalianpur. From this, he deduced by how much a surveyor's plumb line would be deflected from the true vertical at both places. At Kaliana, the plumb would be attracted toward the mountains by 27.853 seconds of arc. Since Kalianpur was farther away, the attraction would only be 11.968 seconds giving a difference of 15.885 seconds of arc.

This was all very elegant, if a little boring, until the good archdeacon looked back at the surveyors' actual results. To his surprise, he found that the true measured variation in deflection of the plumb line between the two places was only 5.23 seconds of arc, less than a third what it should have been by his calculations. Archdeacon Pratt was confident of his calculations and of the obsessive accuracy of the surveyors' work. How then to reconcile the discrepancy? Were the mountains hollow?

Rather than risk venturing out onto the uncertain limb of speculation, the mathematically minded churchman simply wrote up his findings and submitted them to the Royal Society in London. There they caught the eye of George Bedell Airy, an astronomer and the society's president. Airy was not one to shy away from speculation and promptly published "An Hypothesis of Crustal Balance" in the society's *Philosophical Transactions*. Using Pratt's findings, he rather laboriously considered, "the theory of the earth's figure as affected by disturbing causes." It led him to the deduction that "there is nothing surprising in Archdeacon Pratt's conclusion, but that it ought to have been anticipated." In fact, "Instead of a positive attraction of a large mountain mass upon a station at a considerable distance from it, we ought to be prepared to expect no effect whatever, or in some cases even a small negative effect."

Underlying Airy's rather heavy verbiage was a very simple idea. The crust of Earth was less dense than the mantle (Airy envisaged a fluid) below it. If mountains were not simply piles of crustal rock sitting on top of the mantle, but pushed down into it, in the same way that an iceberg pushes down into the water in which it floats, then the mass of a mountain range would be less than expected, since the root below it was actually displacing denser mantle material, and the mountains would therefore be buoyant.

Take a block of wood and put it in a filled bath. It doesn't sit on top of the water but partially sinks into it, displacing some water. Thus, by the top-down density profile as explained with the mall and coffee shop example, a large block of wood creates a less dense area than would have been present if the water were there. So too with mountains. I like to imagine the bewhiskered Airy having a eureka moment with a block of wood and his bath, but I suspect it was more of an intellectual exercise.

The idea that, if you load up Earth's crust, it will push down into the mantle until balance is reached (and, conversely, if you unload the crust, as when an ice cap melts, for example, it will rebound) is called "isostasy." Of course, it is all much more complex than blocks of wood in a bathtub; there are horizontal stresses that, in the case of the Himalayas,

## Airy Isostasy

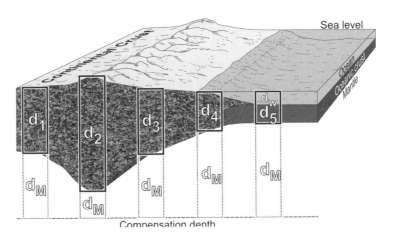

## Pratt Isostasy

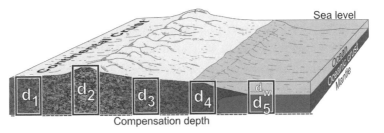

The Airy and Pratt models of isostasy; "d" indicates density. In the Airy model, the depth of compensation is in the mantle, for which the density is shown as dM. The densities in each block, d1, d2, etc., have the same value; the average of the density of water, dw, and of oceanic crust, d5, also has the same value. In the Pratt model, the depth of compensation is at the base of the crust. The densities in each block have different values. The block with the highest topography has the lowest density, d2; the block with the lowest topography has the highest density, which here is the average of dw and d5.

are immense as the Indian subcontinent grinds its way into Asia. There are also several different types of mountain ranges, depending on whether continents are colliding with continents or subducting ocean plates or arcs of volcanic islands, and whether the plates are meeting head-on or sliding past each other at an angle. However, in essence Airy and Pratt (who overcame his reluctance to speculate in 1859) got it right. And the consequences of their insight would prove immensely valuable to the Lithoprobe scientists almost 140 years later.

In a footnote to this story, ten years after Airy first pondered Archdeacon Pratt's findings, and over objections that the mountain should have a local name, Peak XV was named for the Welshman who had been Surveyor General of India while the Great Trigonometrical Survey had been undertaken: Sir George Everest.

---

### Gravity, Isostasy, and Floating Mountains

Isostasy refers to an equilibrium condition, like floating icebergs, between the crust and the mantle, or the lithosphere and the asthenosphere, that slightly molten part of the mantle immediately below the lithosphere. Gravity is the fundamental force involved. In the case of icebergs, the gravitational force that pulls a floating iceberg down is counterbalanced by the buoyancy force that pushes it up. Similarly, a mountain range in the continental crust is buoyed by forces in the mantle. In terms of density, the relatively lighter continental crust projects a root into the denser mantle that allows the high-standing mountain to "float." Ocean basins show the opposite effect. The relatively heavier oceanic crust, which lies below sea level, does not require a root to balance it, so the crust is much thinner. That such a situation as isostatic equilibrium exists in the lithosphere implies that over the long periods of geological time, the "solid" rocks of the mantle had very little strength and behaved like a highly viscous fluid.

At some depth in Earth, known as the "depth of compensation," all pressures become equal, or "lithostatic." At this depth, the mass of one column of the crust is the same as the mass of a different column, regardless of whether the column is in the ocean, on a flat plain, or in a mountain belt. This brings us to the contributions of George Airy and Archdeacon Pratt. Airy assumed that

**Time 1**

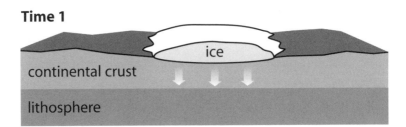

**Time 2**

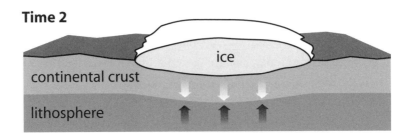

**Time 3**

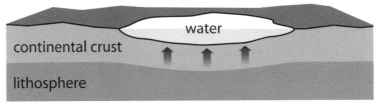

**Time 4**

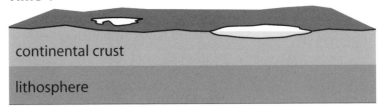

Continental ice sheets, isostasy, and post-glacial uplift. During Time 1, a continental ice sheet starts to form and thickens over the thousands of years of an ice age. Due to the extra load of the ice, the continental crust is depressed; this continues until the buoyant forces from the lithosphere balance the extra load (Time 2). As the ice age ends (Time 3), the ice rapidly melts under the warming conditions, removing the load and causing the depressed crust to rebound upward. Because the rocks of the crust and mantle move ever so slowly, this isostatic rebound continues for a long time after the ice sheet has melted, trying to return the crust to its pre-ice-age level.

crustal blocks have a constant density. Thus the higher the elevation of the crustal block, the deeper it would sink into the mantle. Pratt assumed that the depth of crustal blocks were all the same, that the higher blocks had a smaller density than those at lesser elevations, and if we added up the mass of each of the crustal blocks to the depth of compensation, they all would be the same. In reality, a combination of the hypotheses of Airy and Pratt are most applicable to our studies of Earth.

Isostasy, through a process called isostatic rebound, continues to play a role in the development of Canada. About twelve thousand years ago, most of northern North America was covered with a thick (up to many kilometres) sheet of ice. The centre of the ice sheet was in the vicinity of Hudson Bay. The heavy weight of this ice sheet caused a downward bending of the crust developing a slight depression in the surface of Earth. About ten thousand years ago, the ice sheet melted. As a result the crust began to slowly rebound, just as a foam mattress recovers its form when you remove your body from it. The rebound is shown by raised beaches or terraces around Hudson Bay and it continues to the present day.

---

Mountains, as George Mallory noted before he strode off to meet his end on the slopes of Peak xv, are dramatic and easy to recognize. They rise to spectacular snow-covered peaks that look good on calendars, provide ideal locations for all sorts of extreme sports and act as a magnet for skiers and snow boarders. Unfortunately, they don't last forever. As Bernard Palissy observed in the sixteenth century, mountains are being worn down, grain of sand by grain of sand. Given enough time, mountains will disappear, or rather be redistributed as sediment in deep oceanic basins off continental edges.

One day, after the mechanism operating deep beneath North America has wound down and our continent has stopped its ponderous journey westward, it will be possible to cycle from the west coast to Calgary with little if any appreciable elevation variation. It will be easier on the thigh muscles and less scenic, but whatever evolutionary product of our present world is around to undertake the journey hundreds of millions of years hence will still be crossing a mountain range. The roots will still be there beneath them, not as deep as those beneath the Himalayas, but preserved for future geophysicists to examine.

It is my hope that these yet-to-be-born scientists looking back on our world will not have lost the sense of wonder that draws so many to the Earth Sciences. The question "What was it like?" is a major driving force in the study of history, archaeology and geology. Around the time of my epiphany in the tannery-waste-filled quarry in Scotland, I was asking that question. At that age, my thoughts mainly centred around dinosaurs, movies, and Raquel Welch in a fur bikini, but it was a strong factor in sending me to St. Andrews University to study geology and then to war-torn Zimbabwe to learn about green rocks. In later years, the fact that the work I was engaged in did not address that question was a factor in my leaving geology and taking up researching and writing stories about the past.

Obviously, paleontologists have it easiest. They have concrete objects from the past to deal with and can view accurate reconstructions of vanished worlds in museums and Steven Spielberg movies. However, there is endless pleasure to be derived from continuing Alfred Wegener's work in fitting the continents back together. The supercontinent that Wegener was dealing with, Pangaea and its products, Laurasia and Gondwana, are now fairly well known. After all, their breakup within the last couple of hundred million years is recorded in our own world.

At the other end of the supercontinent spectrum, Ur and Kenorland are still struggling out of the realm of myth. But between these are two other supercontinents that, while still speculative in many respects, are becoming more concrete. The youngest is Rodinia, which formed and broke up between 1.1 billion and 750 million years ago. Prior to that there was Columbia, sometimes called Nuna, which dominated our Earth between 1.8 and 1.5 billion years ago. Spotting and dating the ancient mountain roots that stitched these supercontinents together enables us to unravel their history.

Whether Kenorland was a true supercontinent or not, the world of

2.1 billion years ago consisted of numerous continental fragments drifting in a Pacific-like ocean. Evidence of their interactions over the subsequent three hundred million years can be seen in the relic mountain chains in South America, West and South Africa, Baltica, Greenland, Siberia, India, China, and North America. Slave and Superior (let's give them continental names now) were two of these drifting land masses as were the nearby Hearne, Rae, Nain, Wyoming, Sask, and several smaller fragments.

Centimetre by centimetre, these proto-continents drifted together, crushing oceanic crust between them and thrusting great mountain ranges up along their edges. Hearne and Rae welded together along the enigmatic Snowbird Tectonic Zone, Slave met Rae across the Thelon Mountains and formed the Wopmay Mountains to the west, Nain joined Rae, and Wyoming became attached to Hearne.

Today, these almost unimaginably cataclysmic events have left little to excite the average observer. Deep fractures in Earth's crust, often only understandable through sophisticated geophysics, and narrow bands of highly distorted and faulted rocks, are all that remain of the vast ocean bottoms that separated these islands or the thousands of metres of sediment that washed off their mountains. But one of these sutures binding our world together is different.

Deep beneath the recent sediments of the great plains that generate much of Canada's petroleum wealth lurks the Trans-Hudson Orogen. It appears to be truncated by a later phase of mountain-building below South Dakota, although some postulate its continuance to the west as far as northern Arizona or even northern Australia. In the opposite direction, it extends through Saskatchewan and Manitoba, beneath Hudson Bay, and probably through Labrador and Greenland even reaching into Scandinavia. It is the largest relic mountain belt from the 2.0 to 1.8 billion-year-old growth period of continental formation, remarkable both for its length and for its width.

After its enigmatic journey beneath the American Midwest, the Trans-Hudson Orogen runs beneath the international boundary and finally breaks to the surface in northern Saskatchewan and Manitoba before swinging through ninety degrees to run east between Superior and Hearne. It is here that it achieves its widest development, more than five hundred kilometres in places. It is five hundred kilometres of immensely complex geology—to the untrained eye, a dog's breakfast of different rocks remarkable only in that the swirls of minerals within them are evidence of a long and violent history.

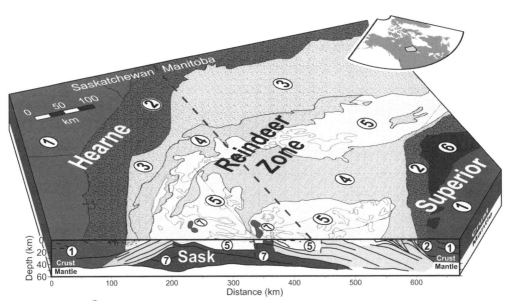

① Hearne and Superior cratons
② Wollaston domain (Hearne) and Thompson boundary zone (Superior)
③ Wathaman batholith
④ Rottenstone and Kisseynew domains
⑤ Glennie, Flin Flon, and Snow Lake domains
⑥ Pikwitonei belt
⑦ Sask craton

The surface of the 3-D block shows a simplified geological map of the Trans-Hudson Orogen in Saskatchewan and Manitoba. Location of the map with respect to northern North America is shown in the inset. The front face of the block is an interpretation of the crustal structure to depths of fifty kilometres based on seismic reflection and geological studies that were carried out across and along the orogen. The numbers on the map identify basic geological units; those on the interpretation show how these units extend below the surface. Note that the Sask craton is isolated from both the Hearne and Superior cratons by thick sequences of other rocks and thus is a separate cratonic block.

## The Trans-Hudson Orogen: A Geological Collage

Lithoprobe studies focused on the Saskatchewan-Manitoba segment of the Trans-Hudson Orogen. Geologists subdivide the orogen into three zones: the Hearne and Superior boundary zones and the Reindeer zone (named after Reindeer Lake, along whose shores the rocks are well exposed).

The Hearne comprises rocks of the Archean Hearne craton (unit 1 on the left side of the map), sedimentary rocks deposited on the margin of the craton (like what is happening off the east coast of North America today), and those associated with the Hearne that were deformed and modified as a result of the collisions (the latter two represented by unit 2 on the left side of the map). The Superior boundary zone has similar rocks (units 1 and 2 on the right side), and also includes some that are known to have formed tens of kilometres deep in the crust but are now at the surface (unit 6).

The Reindeer zone is made up of a complex collage of many different rock types. These include volcanic island arcs (unit 5, similar to the current Philippine Islands); volcanic arc plutons and gneisses (also included in unit 5), the result of island arc rocks melting at greater depths in the crust, then ascending to the surface; sediments in the basins intervening between the arcs (unit 4); sedimentary rocks of the sea floor pushed up and deformed as a result of collisions (also included in unit 4); and continental arc plutons (unit 3), generated by subduction processes and similar to the present Andes Mountains of South America or the Coast Mountains of British Columbia. The places where the Archean Sask craton peeks through the younger rocks are indicated by the two small dark grey areas (unit 7). Together, these rocks provide the clues as to how the complex collage developed.

The simplicity of the Trans-Hudson Orogen map is deceiving because it hides the complexity that is apparent when a smaller area is examined. If we looked at just one part of unit 5, we would find a variety of distinctive rock types, which would be indicative of the complex deformational processes at work in this collision zone. The rock types represent island arcs, oceanic rocks, sediments, volcanic rocks, and plutons that have intruded into all of the other types. These rocks would be separated by large shear zones and broken up into packages by many types of faults associated with the final development of the orogen. The role of the Earth Science detective is to unravel such a complex collage and understand how it was put together. Many of the clues used in this understanding come from geochemical, isotopic, and dating analyses of samples of the different rock types. Such understanding also has a very practical aspect. Scattered throughout the Trans-Hudson orogen are a number of gold and mineral deposits.

Trans-Hudson Orogen seismic studies provide the look below the surface, extending the studies of surface rocks all the way down to the upper mantle. The main seismic profile extends across the entire orogen, so a lithospheric cross-section can be compiled. The simplified section that is shown on the front face of the map summarizes the extent to which different rock domains on the surface extend into the crust. It also shows that the Sask craton (unit 7) is isolated from both the Hearne and Superior cratons and thus is an independent Archean block. Other studies suggest that the Sask craton is a broken-off part from the Wyoming craton, which is found in the United States, to the south and west of our map area.

Yet rock used to be something different. Some formed placidly in deep basins where sediments washed off the approaching continents and collected; others had a fiery birth in arcs of volcanic islands, where crust was already being subducted or where plumes of hot magma rose, as in the Hawaiian Islands; and others are plutonic rocks, released from deep within Earth as super-hot magma rose and solidified along lines of weakness in the tortured crust. The rock types are not unusual in this setting, but why are there so many different kinds?

Two major pieces of North America, Hearne-Rae and Superior, tried to come together where the Trans-Hudson Orogen is widest. Two such vast land masses, driven by the unstoppable forces deep in the mantle, should have crushed and destroyed much of what once lay between them. Himalaya-like mountain ranges should have risen and been eroded to nothing, leaving only a band of tortured rock, perhaps ten to one hundred kilometres wide, above a deep root. The mountains undoubtedly rose and fell, but why did the process stop before the mighty continents came fully together? Lithoprobe found the answer to this mystery.

Australia is moving north at about eight centimetres per year. That means that in approximately twenty-five million years it will crash into eastern Asia. In the process it will pick up all the odds and ends that now lie between, fragments of ancient crust, volcanic arcs and sedimentary basins. The sea floor between all these pieces will be subducted beneath Australia and Asia—Borneo, the Philippines, New Guinea, the Celebes, Ternate, and a host of other islands will be crushed and their rocks deformed almost out of all recognition. History will repeat itself and a new Trans-Hudson Orogen will be born. Until then we only have the original to look at.

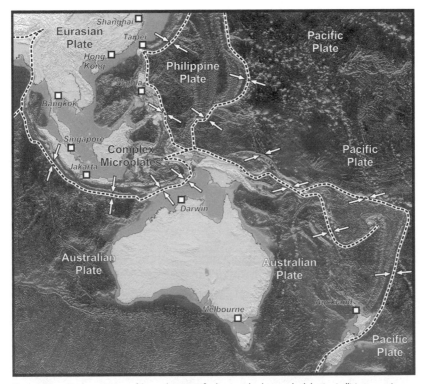

Topographic and bathymetric map of the southwest Pacific showing island arcs and subduction/collision zones (arrows facing each other; their orientation indicates in what direction the plates are moving with respect to each other). The dashed black lines show major plate boundaries. The region between the Australian and Eurasian plates may be similar to the Trans-Hudson Orogen during its development.

Whether Hearne, Rae, Slave, or Superior were once part of Kenorland or not, by 2.1 billion years ago they had drifted apart. Hearne lay in subtropical latitudes while Superior lay chilled close to the pole. Between lay the Pacific-sized Manikewan Ocean, dotted, like the Pacific to the north of Australia today, with small islands.

By 1.9 billion years ago, Rae and Hearne had come together, Superior was approaching, and the stresses on the oceanic plates between were creating arcuate chains of volcanic islands. As the 5,500-kilometre-wide Manikewan Ocean closed, its floor subducted around the margins, forming an ancient "ring of fire" that resembles the one around the Pacific today. By 1.87 billion years ago, many of the volcanic arcs had come together in mid-ocean and were shedding sediment into the surrounding deep water. By 1.85 billion years ago, these arcs had collided with Rae-Hearne, creating a mountain chain very similar to the Andes.

So far so good. This was a story that had repeated, and was going to continue repeating, countless times as the continents broke apart and reformed in their ponderous but unstoppable dance. But this time there was another element to the tale and it was about to make a significant impact.

Hundreds of millions of years before, probably when Kenorland was breaking up, a greater-than-one-hundred-thousand-square-kilometre fragment had broken off Wyoming. It drifted, subject to the currents in the mantle below, until it found its way blocked by the closing Manikewan Ocean. Between 1.840 and 1.805 billion years ago, this mini-continent crushed and deformed the arc volcanic formations in the remnant ocean basin. As it did so, it was in turn trapped between Rae-Hearne and Superior and pushed down beneath the volcanic formations.

But Sask, as this fragment is known today, was not just a piece of oceanic crust that could be squished and destroyed by its larger neighbours. It was a full-fledged piece of thick continental crust, and it stuck. If you place a pillow between a lintel and a closing door, the pillow will compress and the door almost close. If you place a large rock in the pillow, however hard you push, the door will not close nearly as much.

Sask was the rock in the pillow of the Trans-Hudson Orogen. Rae-Hearne and Superior kept pushing, but their energy was dissipated in lateral movement along huge shear zones instead of narrowing the basin. Eventually, by 1.77 billion years ago, things calmed down. A new supercontinent, Columbia, had been formed, and right across the middle lay the scar of the Trans-Hudson Orogen, with Sask stuck in its throat at the widest part.

---

### Crunching Cratons

More than two hundred million years after the Superior Province was assembled, this continental block was far from the Hearne continental block, but how far and in what direction is not known. On the other hand, some evidence suggests that about 2.45 billion years ago the Superior craton may have been adjacent to the ancient Wyoming craton of the western U.S. and perhaps also the Karelia craton that now forms part of northern Scandinavia. Based on the presence of deformed and metamorphosed sediments on the western margin of the Superior continent and the eastern margin of the Hearne continent, earth scientists assume a large ocean separated the two blocks by 2.1 billion years ago. From paleomagnetic studies, this Manikewan Ocean is estimated to be the size of the current Pacific Ocean, about five thousand kilometres across. During the ensuing three hundred million years, the two cratons drifted toward each other. Tracts of oceanic crust were subducted as the slowly colliding continents forced the closure of the Manikewan Ocean, just as the Pacific Ocean will likely be closed within a few hundred million years from now.

By about 1.9 billion years ago, the Manikewan Ocean was dotted with oceanic volcanic islands and island arcs forming complex patterns of subduction with some spreading ridges, just as the current southwest Pacific Ocean is covered with such features. On the northwestern margin of that ocean (using today's coordinates), subduction had started below the Hearne continental margin. Far to the southeast lay the Superior craton, slowly moving northward. By about twenty to thirty million years later, some of the volcanic terranes had amalgamated to form a large complex that shed eroding volcanic sediments into the surrounding seas. The subduction zones were rearranged, and to the

## 1.92-1.88 billion years

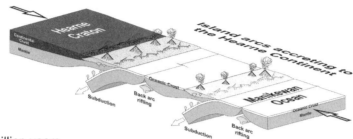

## 1.86-1.85 billion years

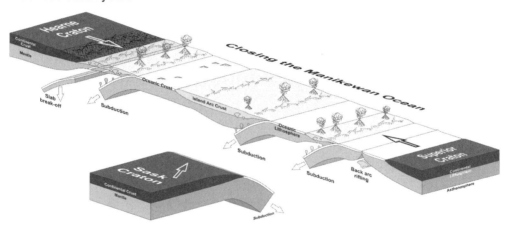

## 1.83-1.805 billion years

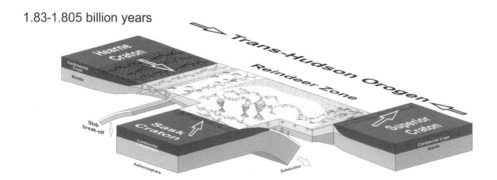

Simplified illustration of three of the many stages in the tectonic development of the Trans-Hudson Orogen. Block arrows show directions of motion. The change in direction of motion of the Superior craton and the initiation of large, crustal penetrating strike-slip faults (thin black arrow) in the lower panel are due to the Sask craton forcing its way into the Reindeer zone, the collision region between the Hearne and Superior cratons.

southwest the oceanic plate on which the Sask craton was riding began to show its presence. To the southeast, a young terrane within the Manikewan Ocean collided with the Superior continental block. In the northwest, subduction and collisions of the strings of volcanic arcs initiated a fold-and-thrust belt, similar to the easternmost Rocky Mountains, in the marginal sediments on the Hearne craton.

By 1.86 to 1.85 billion years ago, the amalgamated accretionary complex in the ocean had thickened considerably due to the continuous crunching. Collision of this complex with the Hearne craton forced subduction of continental and ocean arcs that had formed on the margin of the Hearne continent. The collision and subduction generated a massive new feature, the Wathaman batholith, a characteristic mountain belt like the present-day Andes Mountains in South America. To the southeast, the Superior block retreated a little, leaving some rocks that are characteristic of its retreat. Meanwhile, the plate carrying the Sask craton moved northeastward, bringing the continental block closer to the accreted complex.

Between 1.85 and 1.84 billion years ago, one or more subduction zones within the still-large Manikewan Ocean retreated, allowing formation of an ocean basin. Sediments poured into this basin from the mountain belt, the thickened accretionary complex, and even the Sask craton. By 1.84 billion years ago, the Sask craton had begun its collision with the accretionary complex, causing massive deformation. The collision with the Sask craton, and also possibly the Superior block as it moved northward, caused inversion of the ocean basin rocks.

During the period 1.83 to 1.805 billion years ago, the Sask craton continued to move northeastward, further deforming the accreted complex and forcing it, and the ocean basin rocks, up and over the older block. The Hearne margin continued to deform but was pretty much complete by this time. However, the Superior craton continued its slow northerly march, aiming for a collision with the Hearne. On its way, it collided with the Reindeer zone, thickening the ocean basin sediments and forcing them to also be thrust over the top of the Superior craton.

But the craton-to-craton collision with the Hearne never occurred. The Sask craton provided a solid wedge that prevented the head-on collision, resulting in the preservation of the Reindeer zone, the world's largest such zone of that age. Although the crunching was blocked, further deformation of the now-enlarged continent continued as strike-slip movements along a number of faults led to a southeastward extrusion of the internal parts of the orogen. By 1.77 billion years ago, all was quiet. Today's cross-sectional image has been preserved since that time.

---

Sask lay hidden beneath other rocks, only peeking through to the surface at a couple of places, and giving no hint of its true size and significance, until the geophysicists of Lithoprobe looked below the surface.

Of course, while Rae-Hearne, Superior, and Sask were jockeying for position around the doomed Manikewan Ocean, the rest of the world was going about its own business. In particular, Slave was continuing an already ancient history.

Slave is the grandfather of continents. By the time the mountains of the Trans-Hudson Orogen were rising along the edges of Superior and Rae-Hearne, the origins of Slave were already as far back in time as the Manikewan Ocean is from us. If there were supercontinents before Columbia, then the Acasta Gneiss was a part of them. But it was time for Slave to settle down and, just as Australia's future collision with east Asia is a model for the Trans-Hudson Orogen, there is a modern analogy for what was happening on the other side of Rae-Hearne.

Born in 1869, Princess Maud Charlotte Mary Victoria, the fifth child of Edward, Prince of Wales, and one of Queen Victoria's many grandchildren, was considered one of the best-dressed women of her age, her wardrobe warranting an exhibition all to itself at the Victoria and Albert Museum in London. She was also the first queen of an independent Norway and has several remote and inhospitable parts of the world named for her. There is a Queen Maud mountain range and a Queen Maud Land in Antarctica, a Queen Maud Gulf just south of where Sir John Franklin and his men met their tragic end in the Canadian Arctic, and there is a Queen Maud Uplift sandwiched neatly between the ghostly remnants of the Thelon Mountains, which run along the eastern edge of Slave, and Rae-Hearne. The Queen Maud Uplift is thought to be the eroded remnant of a high plateau created when two continents crashed together. It is roughly triangular in shape, tailing off to the south into the Great Slave Lake Shear Zone, a fundamental suture in Earth's crust.

So, take a map of India and rotate it clockwise through ninety degrees. From left to right, you now have the Indian subcontinent, the Himalayas, the high plateau of Tibet and the Asian continent. If your map extends far enough, you will also have part of Asia at the bottom, wrapped around the Bay of Bengal, across the Sagaing transform, a fundamental suture in Earth's crust.

Now look at a geological map of the area around Slave. From left to right you have the Slave continent, the roots of the Thelon Mountains, the eroded Queen Maud Uplift, and the Rae-Hearne continent. To the south, you have a part of Rae-Hearne wrapped around the Slave and separated from it by the Great Slave Lake Shear Zone.

It is a dangerous thing to draw parallels across two billion years, especially since the scale of the Slave is about one third that of India, but it's a lot of fun, and the similarities are intriguing. After all, it makes sense that similar pieces of crust acted on by similar forces will produce similar results.

The Slave Rae-Hearne collision dates from 1.9 to 2 billion years ago, just at the time that the Manikewan Ocean was beginning to close. Almost as soon as Slave docked with Rae-Hearne, things began happening on the far side of the continent.

The new Slave-Rae-Hearne was assaulted from both sides. As Superior approached from the east across the Manikewan Ocean, another ocean was closing on the other side of the continent. It had no name and there does not appear to have been an equivalent to the Superior crushing from even farther out, but there was enough force and enough islands of volcanic rocks to create a respectable mountain chain.

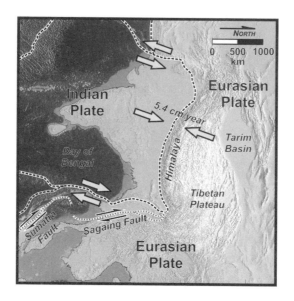

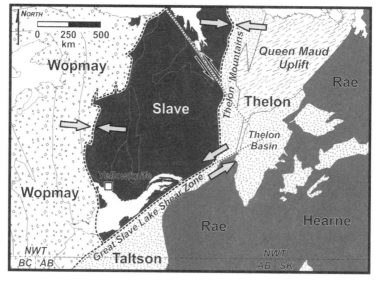

Topographic map of India and area to the north rotated clockwise by ninety degrees (upper panel); and simple tectonic map of the Slave Province and surrounding regions (lower panel). Dashed black lines show major plate boundaries; block arrows show the relative directions of motion between plates. Thin black arrows show relative motions on a few major transcurrent faults (light dashed lines). The "Thelon Mountains" are now eroded remnants. Note the similarity between the two panels, including mountain belts (Himalaya and Thelon), areas of uplift (Tibetan plateau and Queen Maud uplift), basins (Tarim and Thelon) and main plate boundaries.

### Seismic Imaging of Two-Billion-Year-Old Collisions: The Wopmay Mountains

In the winter of 1996, an unusual caravan of vehicles was active along the only road between Yellowknife and Nahanni Butte in the remote Northwest Territories. The caravan contained a seismic crew working for Lithoprobe. They were acquiring reflection data to image the subsurface geology along this 725 km stretch of roadway, from the ancient Slave craton in the east and westward across the Wopmay Orogen to the beginning of the Cordillera. Little did the crew know that they were acquiring one of the best lithospheric seismic profiles ever recorded anywhere. The images generated after computer processing are truly magnificent (at least to a seismologist) and provide considerable new insight into collisions that occurred almost two billion years ago.

By 2.65 billion years ago, the various components of the Archean Slave craton had amalgamated into a continental block. After 2.0 billion years, a part of the block on the west side drifted off, leaving the western part of the Slave which we can now identify (such as the Contwoyto and Anton domains and Yellowknife basin) and space for the formation of the west-facing Coronation passive margin. Between 1.92 and 1.90 billion years ago, a magmatic arc, the Hottah terrane, developed in the ensuing ocean basin, together with the sediments of the margin. As the western Slave craton and Hottah terrane converged, the sediments were compressed, shortened and pushed eastward, until the Hottah completed its collision with the Slave at about 1.88 billion years.

Between 1.88 and 1.84 billion years ago, subduction occurred below the western margin of the amalgamated Hottah-Slave continental block, generating another magmatic arc, now called the Great Bear. Molten rocks from deeper in the lithosphere squeezed their way up to the surface to be deposited on both the Hottah terrane and deformed Coronation margin sediments. The paths of the ascent of these molten rocks are so small that they left no telltale imprint. The character of Lithoprobe's seismic reflection image of the Hottah terrane showed only the thickness of the Great Bear magmatic rocks now overlying Hottah.

Generation of the Great Bear magmatic arc was probably associated with the development, before 1.845 billion years, of the Fort Simpson terrane, an

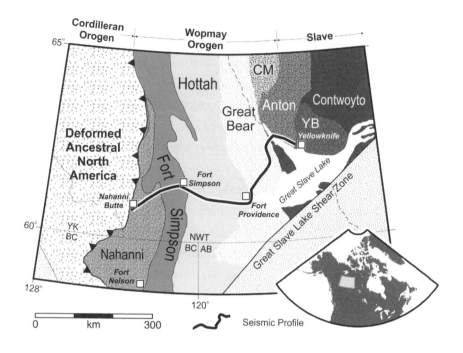

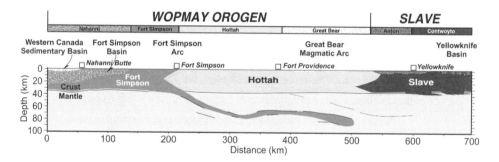

Tectonic map of the Wopmay Orogen and surrounding regions (upper panel); and a simplified cross-section of the crust and mantle along the seismic profile based on interpretation of seismic reflection and refraction data (upper panel). The inset shows the location of the main map within North America. The area on the map between the toothed line, the deformation front of the Canadian Cordillera, and the dashed line is presently covered by sediments of the Western Canada Sedimentary Basin; for the map these have been stripped away. CM is the Coronation Margin and YB is the Yellowknife Basin. The bar at the top of the lower panel shows the tectonic units along the profile. Shading in the cross-section corresponds to that on the map. Light, thin lines in the crust indicate interpreted reflections and show how they define the collisions that took place; for example, the wedgings of Slave into Hottah and Hottah into Fort Simpson. Darker lines in the mantle show reflections from depths as great as 80 km. The darker grey area in the mantle is interpreted as the remnant of the ancient oceanic lithosphere that was subducted during the collision of the Fort Simpson and Hottah terranes and preserved in the mantle for the last 1.8 billion years.

oceanic volcanic arc, in the basin to the west and the convergence of that terrane with the Hottah-Slave block. The timing of this second great collision is not known well, perhaps it was about 1.840 billion years or maybe tens of millions of years later than that. What is known well, thanks to the spectacular Lithoprobe images, is the *style* of that collision, which has been preserved for more than 1.8 billion years. The Fort Simpson terrane is the westernmost known tectonic component of the Wopmay mountain-building episode. Based on magnetic anomaly maps, another component, called Nahanni, has been identified in the subsurface, but because it does not come to the surface anywhere its identity remains an enigma.

For more than a billion years, the Wopmay Orogen formed the western edge of North America. To its west lay an ocean, possibly formed when an unknown micro-continent moved away from Wopmay. On the Wopmay edge of this ocean, stretching of the lithosphere caused by the westward motion of the micro-continent resulted in the formation of a series of very deep basins, up to fifteen kilometres deep, into which sediments were deposited. But this was not entirely a quiescent process. At various time during the billion years, some tectonic deformation caused contraction and uplift of some of the sediments along the margin, as evidenced by geological studies of outcrops and the Lithoprobe reflection data.

---

The Wopmay Mountains rose between 1.95 and 1.84 billion years ago, and consisted of two major pieces of volcanic crust and the sediments that washed off them. They no longer rise majestically. The volcanic rocks and sediments have been distorted and partially melted, but the roots of these mountains are still recognizable and datable.

So, by about 1.8 billion years ago, what we recognize as the North American continent had settled down and was largely assembled. We don't know what shape it was, and can only guess at where its edges were and if they were fringed by long-vanished beaches or welded to other continents, but the Slave, Rae, Hearne, Superior, Nain, Wyoming, and Sask orogens were united to form Laurentia. And this new landmass was important. It was probably the largest continental fragment and formed the heart of a supercontinent. Sometime between 1.8 and 1.5 billion years ago, other fragments that now form Baltica, Siberia, Africa, Australia, India, and South America attached themselves to North America to form the ancient supercontinent Columbia.

It's a great story, but is it true? Is the neat tale of a tidy coagulation of continents and their subsequent clean breakup between 2.1 and 1.5 billion years ago anything more concrete than a mix of intelligent guess-work, inspired speculation, and out-and-out fantasy? Probably, but how *much* more is anyone's guess. Not all the information fits. Paleomag-netism (all those tiny compasses fixed in the cooling lava), uncertain as it is that far back in time, suggests several conclusions: first, that ancient Laurentia could not have been a part of Columbia until sometime after 1.77 billion years ago; second, that Siberia and West Laurentia do not appear to have broken apart until a mere 530 million years ago; third, that some evidence suggests that, if there was a Columbia, it only broke up into two huge pieces before reforming into Rodinia; and fourth, that the Snowbird Tectonic Zone, the place where Rae and Hearne are stuck together, shows evidence of having been active as far back as 3.2 billion years ago and thus could not be the result of Columbia coming together.

Just because plate tectonics was up and running in a recognizable fashion around two billion years ago, this shouldn't lull us into thinking that the world back then was the same as it is now. Uniformitarianism trains us to believe that the rates we see things happening at today have been constant in the past. This is not the case at all. Our early Earth was hammered by levels of meteorite bombardment that would quickly wipe us all out today; the rates at which Palissy and Hutton's grains of sand washed off mountain tops used to be much more dramatic when there was no land vegetation to hold the sediments in place; the ice caps of the last ice age melted and vast floods of melt water carved valleys across the continents. Two billion years ago on any of the continents, nothing walked, swam or flew around, there was no grass or leaves for a cool breeze to blow through, and that breeze was mostly the movement of air made of carbon dioxide, carbon monoxide, sulphur dioxide, chorine, nitrogen, helium, ammonia and methane, but radical change was in that poisonous air. A holocaust was happening in the oceans.

The time around two billion years ago was in the Paleoproterozoic (for the perfectionist, the Rhyacian and Orosirian). Prior to then, in the Neoarchaean, life was dominated by creatures that did not require oxygen to live. This was sensible, as, for most of that time, there was about as much free oxygen on Earth as there is today on Jupiter's sixth moon, Europa; that is, at sea level, enough to fill about half a dozen covered football stadia. As a consequence, the iron parts of a time machine wouldn't rust, although, given the harsh atmosphere, they would probably corrode in other ways pretty quickly. Iron in the ocean was precipitated in layers on the sea floor, but as sulphides rather than oxides.

Then all that changed when, geologically speaking, the oxygen content of the atmosphere shot up. Things began to rust as iron found an element with which it would like to combine. Suddenly, there were no more layers of iron sulphides on the ocean floor, but rusty-red sediments on the land. We owe this extraordinary step forward to one of the lowliest of creatures.

Blue-green algae are now called oxygen-eliminating photosynthetic prokaryotes or cyanobacteria. The new names lose some of the poetry of the old, but they do sound more important, and blue-green algae are very important. Throughout most of the time we have discussed so far, blue-green algae were busy producing oxygen. At first, most of this was taken up chemically, but eventually—blue-green algae are nothing if not persistent—the freed oxygen levels in the atmosphere began to rise. This ancient pollution was an environmental catastrophe for the life forms that had no need for oxygen. Fortunately, some managed to adapt and evolve. And that was not all; at just this time, even the stately progress of the continents seems to have been different.

Between 2.2 and 1.8 billion years ago, new crust was being formed and preserved at a remarkable rate. Small microcontinents, like Hearne and Slave, were becoming the exception rather than the rule, and the cores of the continents we know today were growing and taking shape.

Ancestral North America has been pared down at times and grown at the edges at others, but it has not fundamentally changed since this time.

Why this was happening is still a mystery. Perhaps it had something to do with the mantle cooling to the optimum temperature for plate tectonic processes to occur, or perhaps it was something entirely different. We cannot even be sure which differences we see today are real and which are a result of us only being able to look at part of the picture, but we can say with assurance that things were different two billion years ago.

It's impossible to say how the rates of plate tectonics might have related to the atmospheric changes, but it was an exciting time. As time progressed after this revolution, things transformed in a more or less familiar manner. Partly this is a result of us having more and more history in the rocks preserved for us to look at, but proto-North America, known as Laurentia after about 1.5 billion years ago, was a permanent union. Whatever other pieces of the ancient supercontinent of Columbia came and went, Laurentia stayed together. But it was not always easy. The forces within the mantle were continually restless and they did their best to rip Laurentia apart. They very nearly succeeded in splitting the continent down the middle, but that's another story.

# Unorthodox Fishing
by John Percival

My FIRST MEETING WITH GORDON WEST occurred in a ware-house in Timmins, Ontario. I decided to drop in to visit the head-quarters of the seismic refraction experiment being conducted during the summer of 1984. In those days the seismic instrumentation was primitive: the headquarters had a collection of seismometers of different vintages, makes, circuitries, and states of repair. They came from half a dozen university geophysics departments spread out across the country. I encountered Gordon, whom I knew only by reputation as an esteemed geophysicist, at work with a soldering iron and oscilloscope, trying to coax one more experiment from a recalcitrant recording device. Professor West proceeded to educate me on all aspects of the survey, from the regional scale required to understand the crustal structure, to the circuit scale necessary to acquire the data.

Later during the same experiment I visited Alan Green in nearby Chapleau, Ontario. Alan was in charge of deploying and discharging source explosions for the seismic refraction work—a task that he seemed to relish. The seismic source was hundreds of kilograms of explosives, packed into forty-five-gallon pickle barrels, which were lowered into lakes and then discharged at preset times. On the day I saw Alan, the morning shot had been monitored by the Ontario Ministry of the Environment, which was concerned about environmental damage to the lakes from the explosions. Alan described the simultaneous experiment the ministry had devised to monitor effects of the explosions on the aquatic popula-tion: small fish housed in cages spaced at distances of 50 m, 100 m, 200 m, and 400 m from the source explosion. He then reported on the results of the ecological experiment (they were instantaneous, in contrast to the seismic results, which required months to years of manipulation and massage). Alan admitted that all the fish died. I don't recall the details of how Alan managed to talk his way through continued field operations following this setback, but the seismic refraction experiment was completed. Perhaps the MOE report on the fish experiment should be added to the list of Lithoprobe publications.

## CHAPTER 5

# BREAKING UP?

They say that breaking up is hard to do
Now I know, I know that it's true
—N. Sedaka and H. Greenfield

IN 1698, A SHIP DOCKED IN LONDON with a strange passenger. Called a "pygmy" in an age before everything was neatly catalogued, the creature was in fact a juvenile chimpanzee, and it was dying. Injured in a shipboard accident, the chimp had an infected wound, and only a few weeks to live.

In the late seventeenth century, the world was still large enough to have room in its unknown regions for almost any imaginable wonder, and sailors continually brought back curiosities in hopes of generating a little fame or extra cash. Most curiosities perished on the way or soon after they arrived. This chimp, however, was destined for a posthumous fame that has led to his display today in the Natural History Museum in London.

When this latest curiosity died, someone told Edward Tyson, a local doctor with a passion for dissecting everything he could lay his hands on, from dolphins to rattlesnakes to tapeworms. With the help of illustrator William Cowper, Tyson painstakingly dissected the chimp and compared

149

what he found to other apes and humans. The result was a book, *The Anatomy of a Pygmy Compared with that of a Monkey, an Ape, and a Man,* which T. H. Huxley in 1863 called "a work of remarkable merit" that "served as a model to subsequent inquirers."[11]

Edward Tyson. The anatomy of a Pygmy compared with that of a monkey, an ape and a man ... Second Edition. London : Printed for T. Osborne, 1751. King's College London, Foyle Special Collections Library (St. Thomas's Hospital Historical Medical Collection 14.g.4)

*William Cowper's drawing of Edward Tyson's unfortunate baby chimpanzee, complete with unlikely cane.*

Tyson and Cowper made several mistakes mainly because they did not realize that they were dealing with a juvenile, a fact that undermined several of their comparisons. Nevertheless, *Anatomy of a Pygmy* stands as one of the best early works of comparative anatomy and a foundation stone of primatology. Of course, there was a long and tortuous road to travel before Wallace and Darwin developed the beginnings of the idea that our history should also be considered within the field that Tyson had begun.

Even today, there is considerable controversy among paleontologists studying our own origins, and new finds are revolutionizing thought almost before the last one can be incorporated into a coherent theory. However, everyone agrees that the focus of studying our ancestors has moved strongly into Africa, almost to the environment where Tyson's "pygmy" was born. And our own origins might be closely tied to plate tectonic processes similar to ones that came close to tearing North America apart when our ancestors were little more than blue-green algae.

The discovery of a skull cap and a few bones in a cave in the Neander Valley near Dusseldorf in 1856 led to the first recognition of a member of one of our hominid relatives. Other discoveries in Java, China, and South Africa broadened our knowledge, but it was only with the work of the remarkable Leakey family in and around Olduvai Gorge in Kenya that the motherlode was struck. The dramatic find in 1974 of a new species of hominid in the rich fossil beds of the Afar region of Ethiopia added a new chapter to the story. Paleoanthropologist Donald Johanson's naming of the first specimen "Lucy," after the Beatles' song "Lucy in the Sky with Diamonds," ensured a place for hominid paleontology in popular culture. Discoveries continued and soon there was a bewildering array of skulls and teeth for museums to put on display. But not all the finds were bones.

Around 3.75 million years ago, a volcano erupted in East Africa. It covered the landscape with a layer of ash that became unpleasantly gooey after a short rain squall. Many animals crisscrossed the area in search of food or simply looking for an ash-free environment. Two of the creatures walked on their hind legs. They walked side by side in no great hurry. They weren't hunting or being hunted, just passing through. The couple had a child, who didn't particularly like the wet ash squeezing between his or her toes and so danced along behind in his or her father's footsteps.

These prints in the Laetoli ash, just to the south of Olduvai, prove that our ancestors walked on their hind legs a remarkably long time ago, even before their brains had grown much beyond the size of our ape cousins. Indeed, suggestions of bipedalism go back farther, closer to the time, six million years ago, when we diverged from Tyson's "pygmy's" ancestors.

Six million years ago was an interesting time in East Africa. India had split off tens of million of years previously and was already forcing up the Himalayas and the Tibetan Plateau, but the mantle beneath Africa was not quiet. Rising heat, along a line running from Syria to Mozambique, caused the land to bulge upward into highlands high enough to affect the climate. Rain that would have fallen to the east now fell on the highlands, creating a rain shadow along the coast. As the area dried out, the rain forests died back, creating more open grassland and putting environmental stress on the tree-dwelling primates who lived there.

Eventually, the bulging became too much, and the crust cracked along the top. This formed deep, wide rift valleys that exacerbated the rain shadow effect and contributed to ecosystem fragmentation.

Somewhere, when all this was going on, a small ape that was probably already quite good at upright walking along tree branches began to spend more time in the increasingly open spaces between the trees. Being able to stand upright was an advantage, since it allowed the small ape to see danger coming farther off and to spot possible food sources. Walking became a good thing and evolution did its job. Of course, nothing happens in isolation, and the move to an upright stance had unforeseen consequences.

Walking on your hind legs requires several modifications to the pelvis, spine and skull for balance, strength etc. One consequence of the changes to the pelvis is a narrowing of the birth canal. Thus, young must be born at a relatively early stage in their development rather than grow comfortably and safely *in utero*. A narrow birth canal also means that the head of the infant cannot be very large. So these strange little apes adapted in one way to their changing environment, but they paid a price.

A great advantage for most animals is getting their offspring self-sufficient as quickly as possible. This usually happens in one season and the advantages are obvious—offspring that can fend for themselves are much less of a target for predators, the parents have a greater ability to search for food without being tied to a nest or caring for infants, and the sooner young reach sexual maturity and begin to breed, the better it is for the species as a whole. The length of time an offspring is vulnerable depends on its rate of growth and the amount of development that can take place before birth.

Growth rates vary considerably among infant creatures: reptiles generally grow slowly and throughout their lives while some sauropod dinosaurs grew at extraordinary speed to attain their vast bulk in a dozen or so years. Primates are complex animals and need considerable time to develop sufficiently to function in their world—a baby chimpanzee takes three to four years before it is independent, and young primates are not good at protecting themselves. They make tasty snacks for leopards, hyenas, raptors, and other carnivores, and must rely on living in the trees, dense jungle, or in large groups. Our early bipedal ancestors had to bear their young earlier in their development because of the narrow birth canal, thus increasing the length of time an infant had to be under an adult's care. Simultaneously, they were moving out onto the plains and giving up the protection of the trees.

It must have been a very difficult time, and there were probably points where it was touch and go whether this new adaptation would succeed or fade out as another failed experiment. Fortunately, there were a couple of long-term advantages to being born early and nursed for a long time.

Our lifespan and physical development relative to other species suggests that human gestation should be about twenty-one months. The narrowing of the pelvis restricts this to nine months. After their birth, underdeveloped human infants must be cared for by their parents for three or four times as long as a young chimpanzee before they can venture out and about fully on their own. The advantage of this change was that it necessitated the development of a very strong parental bond and allowed a lot of time for family and other group members to teach the youngster.

Baby chimps can be taught a lot of neat things including basic language and tool use, but there is a limited time when their brains are developing rapidly and they are most receptive to learning. Baby humans have a longer curious phase, which is a distinct advantage, and their brains are bigger.

A chimpanzee brain weighs about seven ounces at birth. Over the first few years it grows to about fourteen ounces in an adult. A human brain weighs about fourteen ounces at birth, way smaller than it should be, but you can't have a big head and a small birth canal. In the first year an infant's brain grows to thirty-five ounces. Continued growth adds another ten ounces before the brain reaches its full size at around five or six years of age.

Our brains start off bigger than our closest primate relatives, and they grow much more before maturity. It is possible that combining a long, protected childhood with a large brain that grows phenomenally after birth when it is subject to all the many stimuli of the outside world was a very important factor in the development of our intelligence. Compared to a chimpanzee, we have longer to learn and more to learn with, all because we walk upright. Bipedalism, which developed as a response to climate and environmental change caused by uplifting and rifting of Africa, led us to where we are now. Once again, plate tectonics is our parent.

Of course, the road wasn't smooth. Over the millennia, in this five-thousand-kilometre African valley that seems to have been so extraordinarily well suited to our development, nature tried many evolutionary adaptations to the changes that were happening. Eventually, one hundred thousand to two hundred thousand years ago, these changes led to an incredibly successful species that exploded out of the valley and conquered the world. At least, that seems to be a reasonable theory until or unless we make some new discoveries that force us to do a rethink.

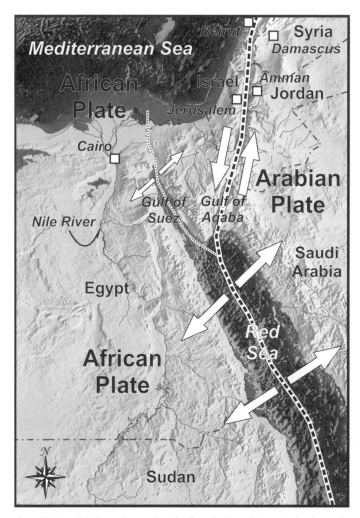

The Red Sea is a relatively new divergent boundary formed as the Arabian peninsula split away from Africa. This splitting also opened great rifts, including the Gulf of Suez and the Gulf of Aqaba, now flooded by the sea. The black dashed line is the plate boundary and shows the divergent boundary in the Red Sea transforming into a transcurrent (or transform) fault boundary through the Gulf of Aqaba and further north. The arrows show directions of plate motions.

The Great Rift Valley in East Africa allowed us to evolve and succeed spectacularly; however, in many other respects it was probably a colossal failure.

Continental rifting occurs in four stages. The first is the rift valley formation such as the one that created the environment in East Africa that our ancestors seemed to like so much. The valleys formed in this way vary considerably, ranging from thirty to one hundred kilometres wide and from a few hundred to several thousand metres deep. The valley edges tend to be steep and faulted and volcanic activity is common.

In the second stage, if the mantle hot spot beneath the rifting develops, the two sides of the valley pull apart and the floor continues to subside. Eventually, sea water floods in and a narrow arm of the ocean forms. This has happened to the northern section of the African rift system to form the Red Sea (see illustration on page 72).

At the third stage, as the pieces of continent move apart, lava upwells to form new oceanic crust. This is beginning to happen in the Red Sea.

By the fourth stage, a true ocean, like the Atlantic, is formed. As the oceanic crust moves away from the hot mantle beneath the mid-oceanic ridge, it cools and subsides, forming basins that fill with sediment washed off the continents. Eventually, if the movement of the continents is reversed, these margins become the sites of subduction zones as the ocean is destroyed.

But the rifts where we came down from the trees are not the only examples in Africa. There is a rift complex on the west coast, too. It is much older, dating from about 120 million years ago, and is important today as the site of much economic exploration in a tragically unstable political region. A major component of this system is the Benue Trough, running for one hundred kilometres from the Niger Delta to beneath Lake Chad. It progressed farther than the east African rifts, developing as a narrow seaway before filling up with volcanic rocks and sediments.

The interesting thing about the Benue Trough is that it is the same age as the successful rifting that began the opening of the Atlantic Ocean. Oddly, it runs at almost ninety degrees to that successful rifting. How can rifting that produces a linear feature like an ocean also produce an arm at right angles?

Easy, if you assume that the hot spots below the domed and rifting continents are not regular linear features. We know hot spots exist. The classic case was studied by John Tuzo Wilson and is currently forming the Hawaiian Islands. This hot spot has been active for millions of years as the Pacific Plate has moved over it. Thus it has formed a line of islands, getting progressively older and more dormant to the northwest of the Big Island of Hawaii. What if, instead of a single, discrete hot spot, they formed a chain of several hot spots, close enough together to rip the continent apart, but far enough away that they develop independently? It's perfectly possible. After all, it's extremely unlikely that heat flow from the core through the mantle to the crust is nice and regular and organized. It's much more likely that there are plumes of different strengths that sometimes occur singly, or in clusters, or in lines, rather like the way plumes burst out from the surface of the Sun.

So, let's say that there is a line of hot spots beneath a continent, each trying to form a dome. A dome will fracture at the top into rifts just as a ridge will, but not in the same way. A dome will fracture into what is called a "triple junction," three cracks at about 120 degrees apart. If left to itself, all three cracks might form narrow oceans, but what is more likely is that only two, influenced by other domes on either side, will develop and the third will fail. Thus you have an ocean forming with a failed rift running off it at a high angle, exactly the situation with the Benue Trough and the Atlantic.

Donald Johanson's Lucy and her tribe didn't know it, but they walked upright over another triple junction. The Red Sea and the Gulf of Aden form a wide V-shape around the bottom of the Arabian Peninsula. Both marine arms are widening as the floor beneath spreads and Arabia drifts away from Africa. At the base of the V-shape is Lucy's home in the Afar region and, running south, through Olduvai and into Mozambique, is the Great Rift Valley, the third, failed, arm of the triple junction.

No one knows whether the Great Rift Valley will remain a failure and become fossilized, like the Benue Trough, or reactivate and allow east Africa to split off and drift out into the Indian Ocean, much as Madagascar did millions of years ago. What is certain is that the rift valley is still active and might remain so for a long time.

Generally, if you live away from an active continental margin, you don't need earthquake insurance. However, there are exceptions. In the eight weeks between December 16, 1811, and February 11, 1812, the small frontier town of New Madrid and other communities along the Mississippi River were rocked by no fewer than sixteen major earthquakes. The three most dramatic quakes measured close to eight on the Richter scale and rank as some of the strongest earthquakes in recorded history. Trees were thrown around, water spouts shot up ten feet into the air, and the Mississippi River was diverted. If it happened today, with the increased modern population, it would be classified as a natural disaster on an almost unprecedented scale.

Major earthquakes are relatively common where plates come together, for example, along the west coast of North America, the Himalayas and Indonesia, but New Madrid lies in Kansas, near the border with Tennessee. It's supposed to be stable, thick, continental crust, but it wasn't always so.

Around five hundred million years ago an ocean, Iapetus, opened to the south and east of New Madrid. For a while, it was a respectable ocean, several thousand kilometres across and, of course, it began with rifting and a triple junction. Iapetus is long gone, but the failed arm, the Reetfoot Rift system, runs along the Mississippi River beneath New Madrid.

It should therefore not be a surprise if, hundreds of millions of years from now, the Great Rift Valley, wherever it may then be on the face of our ever-changing world, is still influencing the creatures that will be living above it. Hundreds of millions of years in the future perhaps, but not billions. A billion years from now, the Great Rift Valley will be dead. One point one billion years ago, currently dead rift valleys were active, and that is when the next chapter in the Lithoprobe story unfolds.

Had you, Toronto, and Chicago been around 1.1 billion years ago, and had you decided to travel from one to the other, it would have been an interesting journey. It would have been through the heart of the supercontinent, Rodinia. Columbia had broken up four hundred million years before, but the newly welded mass that would become Laurentia had stayed together, drifted and become the core of Rodinia. To the east, today's east, lay something that was not yet South America and to the west lay a proto-Australia and a proto-Antarctica.

It was a strange, reddish world back then, barren and violent. Even this long after the Paleoproterozoic Era, still no animals crawled, slithered, hopped, or walked over the landscape, and no trees towered, bushes flowered, or grasses waved. The atmosphere was not as poisonous as it had been when the Acasta Gneiss formed, but it was still low in oxygen compared to today and you would have been continually breathless. There was also no ozone layer, so you would have required an extremely heavy duty sunscreen. The landscape would have been incredibly rugged, with high rates of erosion and periodic catastrophic flooding.

Given all that, the journey would have been manageable to begin with as you trudged over the rocky ground. It would only be when you came to the escarpment that your real problems would have begun. In front of you, a steep scree- and rock-covered slope would have plunged down several thousand metres to a valley floor dotted with lakes, steaming lava flows and the cones of smoking volcanoes. Perhaps in the far distance you might have glimpsed the darker shadow of the equally daunting far valley wall. If you had any sense, you would turn around and go home, wait a few million years until the valley filled with sea water, or a billion years until the valley was so buried that only Lithoprobe could see it.

What you would have been peering down into was the Keweenawan Rift, a structure similar to the Great Rift Valley of Africa, but (in some dimensions) on a larger scale. The Keweenawan Rift is two thousand kilometres long, running from Kansas north to beneath Lake Superior, where it turns sharply back down under central Michigan. Its exact extent is doubtful, but it has been broadly known from geophysical evidence for a long time. It appears as a succession of basins, the deepest and widest being where the rift changes direction beneath Lake Superior.

---

### The Attempted Break-up of North America
The arcuate Keweenawan Rift, located within the mid-continent of North

America, is generally defined on gravity and magnetic maps and has been known for decades. On the gravity map, it is shown clearly by an arcuate pattern of generally high gravity values against a background of lower values. This was always a puzzle until the Lithoprobe results were obtained. On the magnetic map over Lake Superior and the area to the north, the rift also appears as an arcuate structure. However, it is defined by much smoother magnetic values (or, to a scientist, longer wavelengths) than the area to the north where the Canadian Shield is found. This is because the volcanic rocks within the rift, which carry little magnetic grains, are covered with thick sediments, which are not magnetic. Consequently, the volcanic rocks are much further from the surface than rocks in the exposed shield to the north. For measurements at the surface, this causes longer wavelength anomalies.

The images generated from the seismic reflection data recorded in Lake Superior and Lake Michigan (one example from Lake Superior is shown) showed how thin the crust had become under the lakes. The crust was stretched so much that in some places it was less than ten kilometres thick, about one-quarter of its normal thickness.

The images also clearly showed the incredible thickness of interlayered volcanic rocks and sediments that formed in the rift basin depression. As the volcanic matter was spewing out from below the great rift, it formed layer after layer, similar to the layers from multiple eruptions of a volcano like Kilauea in Hawaii. Later, the volcanic layers were interleaved with sediments pouring into the depression as a result of erosion from the surrounding land areas. At their thickest in the centre of the basin, these rift deposits achieved a depth extent of almost thirty-five kilometres. They thinned toward the edges of the rift, much as you would expect. Images across different parts of the rift also demonstrated that the characteristics of the rift varied along its length. To answer the puzzle noted above, the thickness and high density of the volcanic materials, even with interlayered sediments, generates the relatively high gravity anomaly values.

After the rifting and its associated outpouring of volcanic material ceased, the great depression filled with younger erosional material, clastic sediments up to ten kilometres thick. Along much of its extent, the Keweenawan Rift is now covered with even younger rocks and can only be inferred from the gravity

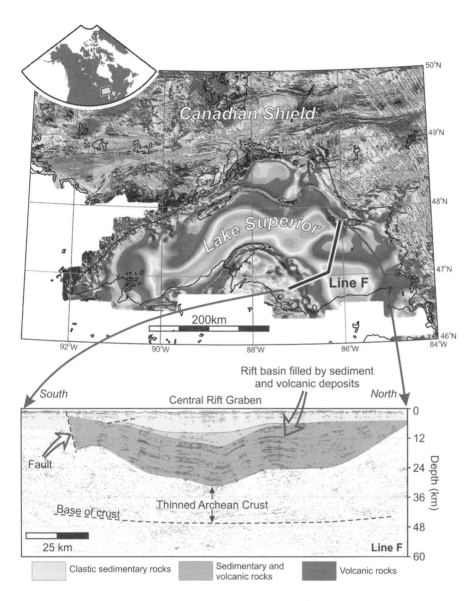

Magnetic anomaly map over Lake Superior and the Canadian Shield to its north; the inset shows the location of the main map within North America. Such maps can be used to help prepare geological maps of different rock types because different rocks have differing amounts of magnetite in them and thus produce distinctive signatures. For example, intrusive mafic dykes usually are more magnetic than the rocks into which they intrude. The linear striations in the Canadian Shield on the right side of the map are caused by the intrusion of hundreds of dykes. The thick solid line labelled Line F is the location of the seismic profile for which the data and its interpretation are shown below. The thin lines throughout the cross-section show the seismic data after much computer processing. The interpretation, shading, and other symbols, are overlaid. In the central rift graben, crust that was probably forty kilometres thick was stretched and thinned to almost ten kilometres. The depression that was formed was filled with volcanic rocks, sediments, and now the waters of Lake Superior; the continuous reflection about one millimetre below the zero line is the lake bottom.

and magnetic data. Along the edges of Lake Superior, remnants of the vast volcanic outpourings are found and have been extensively studied. Such studies provide the ages of the rocks and indicate the subsequent deformation that occurred in the region.

The thin crust and the huge volume of volcanic material that poured out from the rift indicate that what we now call the continent of North America almost split asunder about one billion years ago. The reasons why this rifting was arrested remains a puzzle for Earth scientists to solve.

---

Around 1.109 billion years ago, a hot spot formed in the mantle beneath Rodinia. The crust was pushed up and fractured into a triple junction. The southwest and southeast arms developed and a rift valley formed. The northerly arm failed and formed what is today known as the Nipigon Embayment.

The rift valley floor subsided and lava poured out, much of it beneath the waters that flooded into the depression. For twenty-five million years this continued, thinning the continental crust to only a quarter of its original thickness and building up an incredible thirty-five kilometres of volcanic rocks interlayered with eroded sediments that poured into the deepening rift. The last lavas are the same as those at mid-oceanic ridges, so the Keweenawan Rift came very close to becoming a true ocean and tearing Rodinia apart. But, for some reason, it didn't. The forces pulling the two sides apart stopped, the lava flows dried up, and the huge rift valley filled with younger sediment until nothing on the surface remained.

The geophysicists of Lithoprobe have given us a much clearer picture of the Keweenawan Rift, but we still have no idea why it formed and why it stopped. A look at what was going on nearby at the same time does not make the picture any clearer.

Rodinia stayed together for almost 350 million years. It stretched from pole to pole with Laurentia lying on its side just south of the equator and welded to South America (Amazonia) below it. That weld must have been impressive.

Today, it is possible to climb to the highest point on the planet without the aid of bottled oxygen. It's cheerfully called the Death Zone and you need to be fit and lucky to get there and you can't dally long to enjoy the view. A billion years ago, the highest point on the planet might have been only a few hours' drive northwest of Toronto and no one could have got to it for two reasons—there was less oxygen in the atmosphere to begin with and the peak was probably higher than Everest.

No human ever saw them, but the Grenville Mountains must have been awesome. They ran from Mexico, through Texas, the Midwestern United States, Ontario, Quebec, Labrador, and into southern Sweden. They formed the join between Laurentia, Amazonia, Baltica, and bits of what would become Africa. Similar mountains on the other side joined Laurentia to the rest of Africa, India, Antarctica, Australia, and Siberia.

Today it is possible to climb the Grenville Mountains. But perhaps climb is the wrong word; a gentle Sunday afternoon stroll up from the shores of Georgian Bay off Lake Huron will take you to the roots of those once majestic summits.

The Grenville Mountains rose dramatically, but they fell dramatically too. With no vegetation and a harsh climate, landslides, debris flows, and raging whitewater torrents would have been common, carting huge quantities of rock and sand out onto the surrounding Rodinia plains. In fact, debris from the Grenvilles can be found thousands of kilometres away in the present-day Arctic. As with so many ancient mountains, all that is left is the core: rocks that were once far below the surface and which have been pressure-cooked into almost unrecognizable new rocks. But, with Lithoprobe's help, some of the history of the Grenvilles can be established.

After the breakup of Columbia, Laurentia was not quiet. As it drifted, it picked up debris scattered across the oceans of that distant world. Continental fragments and volcanic arcs were attached to the edges during the long journey. Around 1.45 billion years ago, a volcanic arc similar to the one presently along the edge of the Andes formed along the Superior edge of Laurentia. As this was happening, oceanic plates were grinding past and beneath one another far out to sea. These formed large,

©Galen Rowell/CORBIS

*Not quite the Grenvilles, but the best we can do today—Mount Everest*

arcuate chains of islands with numerous active volcanoes and shallow basins where the detritus from the volcanoes collected and carbonates formed. These islands crashed together, forming ever-larger masses until around 1.24 billion years ago, there were two large belts of new crust.

Inevitably, by 1.16 billion years ago, these two amalgamated to form a significant landmass, sandwiched between Laurentia and the fast approaching Amazonia. The Grenville Mountains proper, consisting of the volcanic arcs being deformed and pushed up onto Laurentia rose between 1.12 and 0.98 billion years ago.

### Georgian Bay and the Roots of Grenville Mountains

By about one billion years ago, the Grenville Mountains (or orogen) had fully formed. And what a spectacle they must have been! Extending northeastward all the way from Mexico to southern Sweden (this was before the Atlantic and predecessor oceans even existed), an ancient visitor from space would see the Grenville Mountains as one of the planet's major features, just as today's astronauts see the Himalayas. Now the ancient mountains as a major topographic feature are gone. Much of their remains are covered by younger rocks, but in Canada, from southeastern Ontario to southern Labrador, some of the remnants of this vast mountain chain are preserved. Lithoprobe seismic images to depths of fifty kilometres reveal the expressions of the tectonic forces at work.

The Grenville mountain-building episode was a lengthy process, lasting for more than 250 million years. During its early stages (about 1.25 billion years ago), micro-continental blocks were progressively stacked on and against the southeast margin of Laurentia, depressing the margin rocks to lower crustal depths. These blocks were transported (in today's coordinates) from southeast to northwest. The seismic reflection data reveal at least one block, probably six to ten kilometres thick, that was thrust up and over other rocks. During a later stage, there was some action near the stable craton. Thrusting action, again to the northwest, cut deep into the crust and possibly the lower mantle, causing rocks that were buried by the earlier stacking to be brought near the surface. The deep thrusting was done under such high pressures and temperatures that the blocks of rocks were somewhat ductile, like Plasticine. Their movement in such a state probably caused the characteristics that generated the strong reflections observed on Lithoprobe's data. Through these processes, the thickness of the crust became very great. Since present thickness is up to fifty kilometres and geological studies indicate that about twenty kilometres have been removed, the original thickness must have been about seventy kilometres, similar to that of the Himalayas. In the ensuing billion years since the Grenville Mountains were a dominant feature on Earth, the high mountains eroded. Remnants of this erosion have been found in the Arctic Ocean three thousand kilometres north of the Grenvilles, highlighting the extent of erosion through mighty rivers and the lofty heights of the mountains. Only the roots of the mountains, as imaged by Lithoprobe, remain.

Lithoprobe

*Rocks at Georgian Bay—the roots of the Grenville Mountains*

So there, at least in broad brush strokes, is a simple story of crust forming and a supercontinent knitting together. But there is a wrinkle. Continents coming together are compressional events. The Grenville Mountains
rose to such great heights because there was nowhere else for the rocks
squished between Laurentia and Amazonia to go but up. Yet, between 1.109
and 1.094 billion years ago, only a few hundred kilometres away, Rodinia
was busily trying to tear itself apart along the Keweenawan Rift.

There are two possible explanations. The coming together of Laurentia and Amazonia was not as simple as it seemed. Perhaps the forces
in the mantle were much more complex than we think and there were
spells of the continents being pushed together and pulled apart. This is
supported by a gap in the data from the Grenville Mountains themselves.
Prior to 1.109 billion years ago, metamorphic minerals were being
formed beneath the mountains, indicating that pressures and tempera-

tures were high. After 1.094 billion years ago, there is also an abundance of metamorphic minerals, which indicate violent plate interaction. However, there is a gap in the ages of metamorphic minerals exactly when the Keweenawan Rift was forming. This suggests that there might have been a realignment of the plates at this time, reducing the stress or even causing some pulling apart. This would explain why the rifting failed.

An alternative theory is that the rifting was due less to extensional forces than heating from below. If a large mantle plume formed beneath Keweenawan, it could have eroded the base of Earth's crust, in the same way that a rising plume of warm water will thin a layer of ice on the surface. The crust domed and cracked, and hot magma forced its way up through the cracks to fill in the valleys where the crust was thinner. The rift occurred because the crust was eroded from beneath rather than pulled apart.

A modern analogy that takes us back close to Lucy's homeland might provide clues. As we have already seen, Africa is busily trying to tear itself apart along the Red Sea, Gulf of Aden and the Great Rift Valley. A mere few hundred kilometres to the west, it is happily crashing into southern Europe, intent upon destroying the Mediterranean Sea. Extension and compression are happening side by side.

The lesson to be learned is that simple explanations, however attractive and comprehensive they seem on the surface, are probably never the whole story when we are dealing with fragmentary information from far back in the geological record. Fortunately, as we move on in time, less of the data has been destroyed and we can speculate on what was happening with a greater degree of certainty.

# Bear Attack
by John Wilson

ONE OF THE ADVANTAGES of doing field work is that you get chances to go places that would otherwise be inaccessible or expensive to get to. The disadvantage is that you are often in places where you come in contact with wildlife. Sometimes both coincide, and occasionally with tragic results.

There is a school of thought that blames bear attacks exclusively on human stupidity. That is certainly a factor and many attacks result from inexperienced visitors to bear country not knowing what to do when they meet one of the neighbourhood's inhabitants. Black bears are generally timid around people and will amble off when a confrontation takes place. However, some black bears, for some reason, become predatory and will attack viciously for no apparent reason. There was one of those at Liard River Hot Springs Provincial Park in northern British Columbia when Arie van der Velden, a twenty-eight-year-old geophysics research assistant from the University of Calgary, stopped there for a relaxing soak on August 14, 1997. He was one member of a large crew that ventured into northern British Columbia to carry out a seismic refraction experiment for Lithoprobe.

Arie was soaking in the pleasant, warm waters of the secluded upper pool of the hot springs when he and the other bathers heard shouts from through the bush. Going to investigate, they walked along a boardwalk towards the well-known Hanging Gardens attraction. Almost at the turnoff, someone yelled that there was a bear and to run. The group turned and fled, but Arie slipped on the wet boards and fell off the path. Instantly, he went from being in a safe, ordered tourist attraction to being in chaotic wilderness with a vicious predator.

Without hesitation, the bear attacked, slashing long tears in Arie's flesh. Arie fought back desperately, pulling the bear's ears and striking it on the nose. Nothing worked. The bear's weight overpowered him and he felt its claws and teeth working on him. The bear pinned Arie against a log and began feeding on his left thigh. He was about to die a hideous death.

The first bullet caught the bear in the neck and probably killed it. The next two were just to make sure.

Arie was extraordinarily lucky. His was the last chapter in a tragic saga that day. By the time Arie fell, the predatory bear had already killed two people, Patti McConnell, a young mother on her way from Texas to Alaska, and Ray Kitchen, a trucker who had gone to Patti's aid and fought the bear, which had severely mauled Patti's thirteen-year-old son, Kelly. More people didn't die because Alaskan businessman Dave Webb had a 30-30 rifle in his camper and Duane Eggebroten knew how to use it. Even when you think you are safe, the hazards of wilderness field work can catch up to you.

# PART 4

## MID-LIFE CRISIS: THE PAST BILLION YEARS

## CHAPTER 6

# FALSE STARTS

Some say the world will end in fire,
Some say in ice.
From what I've tasted of desire
I hold with those who favor fire.
But if it had to perish twice,
I think I know enough of hate
To say that for destruction ice
Is also great
And would suffice.

—"Fire and Ice"
Robert Frost

MY MEMORIES OF FAMILY HOLIDAYS in the west of Scotland are dominated by images of being huddled on a picnic blanket eating hard-boiled eggs and tuna sandwiches, sitting in window seats in drafty bed-and-breakfasts watching the raindrops race each other down the pane, and walking, hunched over against the wind, along rattling pebble beaches.

My mother's favourite word from this time was "bracing," as in: there I am, foaming rollers crashing on the shingle at my feet, dressed in every stitch of clothing I possess, bent double against a force eight

North Atlantic gale, begging to go home, and I am told, "Don't be silly, it's bracing." I have never understood what she meant by that. Oddly enough, it was the opposite of what I was told at home, "Put a jacket on, you'll catch your death of cold out there." Of the many meanings of "brace" in the *Concise Oxford*, I assume she took it as meaning "invigorate," but why freezing half to death should be a bad thing at home and a good thing on holiday always escaped me.

Despite being braced to excess, I enjoyed these holidays. I learned to fly-fish. I came to appreciate the joys of long walks on beaches and over moors, and I discovered rocks and fossils much different from the ones I chipped away at around my home. Our holiday to Ballantrae in Ayrshire in the summer of 1965 was memorable for other than climatic reasons.

Alexander "Sawney" Bean was a caring paterfamilias. Born near Edinburgh sometime in the early fifteenth century, Sawney apparently married a witch, Black Agnes Douglas. This offended the locals who drove them away. Eventually the wandering Sawney and Agnes arrived at Bannane Head near Ballantrae. There they discovered a large cave in the cliffs and decided to set up house. For twenty-five years, so the story goes, despite the cave mouth being cut off at high tide, the Beans struggled to raise a family.

Disinclined by nature to undertake honest work, Sawney and Agnes took to a life of crime. Venturing out of their rather dank home at night, they waylaid solitary travellers, murdered them, and stole their money. As time went on, the family grew and ultimately, eight sons, six daughters, eighteen grandsons and fourteen granddaughters shared the cave in a web of relationships that I hesitate to think too hard about. The growing family presented certain practical problems, not least of which was how to purchase enough food with their ill-gotten gains without attracting too much unwelcome attention. Caring father that he was, Sawney overcame this difficulty in an ingenious way: the Beans would eat their victims, pickling what they couldn't manage at one sitting. This, as it were, killed two birds with one stone, solving the food crisis, while conveniently disposing of the evidence.

Of course, the people in the surrounding area became suspicious as the numbers of disappearing travellers mounted into the hundreds and partly dismembered body parts washed up all along the coast. Searches were mounted, but the cave remained hidden. In frustration, the locals lynched various innocent strangers without much effect.

Things were doomed to go wrong for the Beans' unorthodox lifestyle. One night, the family surrounded a couple returning from a local fair. They murdered the woman, but the husband put up unexpectedly vigorous resistance, holding the family off with his sword until help arrived. Four hundred men and bloodhounds soon discovered the Bean's happy home and led the forty-eight family members away in chains while they tried to prevent their shocked minds dwelling on the sights they had seen in the cave.

Sawney and the male members of his clan were killed by having their hands and feet cut off and being allowed to bleed to death in front of the womenfolk. The women were then burned alive.

Serious historians have cast doubt on the Sawney Bean story, not least because there are no written records and it does stretch credulity to imagine a family of forty-eight living undetected in a cave for a quarter century while they happily murdered and ate over a thousand people. Of course, that didn't bother my fourteen-year-old mind and I shuddered at the thought of Sawney's exploits on every windy expedition to collect rocks and fossils along the coast. It apparently also didn't bother movie director Wes Craven, who used the tale as the basis for his 1977 horror film, *The Hills Have Eyes*. Craven revelled in the cannibalism, incest, and violence, but balked at the bracing weather and set his version in the California desert.

Being made of sterner stuff, I continued my explorations along the Ayrshire coast, although I didn't go into many caves. I took home a box of rocks that soon overwhelmed my gruesome interest in medieval Scottish culinary habits. The rocks were different from any I had collected before. The fossils were not just slightly older or slightly younger than the ones from the quarry at Bridge of Weir, they were fundamentally different.

There were recognizable shells, although they looked different from the ones in the tannery quarry. There were fragments of creatures that looked like the wood bugs that scuttled out of dank, dark corners in the garden, and there were lines on black slate that looked as if they had been drawn in pencil. In addition, the black volcanic rocks that I also collected were like nothing I had seen before. What had happened in this windswept corner of southern Scotland before even the time of the Sawney Bean family?

At home, I cleaned the fossils, dug up my dog-eared copy of the British Museum of Natural History's *British Palaeozoic Fossils* (second edition) and began searching. The fossils were indeed different. They were from the Ordovician era (490 to 443 million years ago), more than one hundred million years before the rest of my collection. The pencil marks turned out to be graptolites, odd creatures that swam or floated around ancient oceans. What had looked like wood bugs to me were actually trilobites, an extinct branch of life that dominated the floors of early oceans.

*The inhabitant of a five-hundred-million-year-old sea. A trilobite* (Elarthis kingi) *from Springs Quarry, Utah.*

The black rocks also proved to be unusual. I couldn't identify them from my books, so I took them along to the local museum. There I was told they were bits of North America left behind when the continents split apart. I created an image of a dark North America, made up of black rocks. It sounded less likely than Sawney Bean's story to me, but I filed it in my memory.

Over the years, I learned about continental drift and many other wonders from reading Arthur Holmes' delightful book *Principles of Physical Geology*. Then I went to university and discovered plate tectonics and the true origin of those black rocks from Ballantrae. They weren't a discarded part of North America at all, but rather bits of a long-vanished ocean—and it all led to another tale of incest and cannibalism.

In Greek mythology, Iapetus was a first-generation Titan, a son of Uranus and Gaia (Heaven and Earth) and father, by his niece, of Prometheus, Atlas, Epimetheus, and Menoetius. Iapetus' brother Cronos, afraid that his own children would depose him, ate them. He missed Zeus, being tricked into eating a rock instead, and so began the clash of the Titans. Zeus won and the rebel Titans were cast into gloomy Tartarus.

Iapetus' kids didn't fare particularly well even after they escaped their father's culinary habits. Atlas fought on the side of the older Titans and was given the heavens to carry when they lost. He was turned to stone for looking at Medusa's head and now forms the Atlas Mountains in Africa. Menoetius was rude to Zeus and smitten by a lightning bolt. Prometheus and Epimetheus supported Zeus and were given the task of creating creatures to populate the world. Both set about their task, using the gifts Zeus had provided for sharing out.

Epimetheus was the Type A personality and worked night and day to create the animals, distributing the gifts—claws, courage, speed, feathers, fur, and so on—with wild abandon. When Prometheus finished carefully making his contribution, a pale, weak copy of the gods themselves, there were no gifts left. To compensate, Prometheus stole fire from Zeus as a gift for his creation. In punishment, he was chained to a rock in the Caucasus Mountains and doomed to have an eagle eternally eat out his liver. Epimetheus didn't escape, being destined to marry Pandora, whose curiosity let all humankind's woes escape from her box, leaving us with only Hope.

So Iapetus, through his two rather incautious sons, is an ancestor of humankind. By way of thanks, astronomer Giovanni Domenico Cassini named one of Saturn's stranger moons after him in 1671 and, more recently, the ocean whose remains formed Sawney's cave was called Iapetus. The Iapetus Ocean may not be an ancestor of Homo sapiens, but it was forming and being destroyed as our distant ancestors discovered the benefits of a shell or a skeleton and one particular group began experimenting with a long cord running down the back that would one day become a spine.

The story of the Iapetus Ocean is familiar—rifting, flooding of a narrow seaway, formation of ocean crust, island arcs growing and then crashing into continents, subduction and the final rising of a huge mountain chain as the continents came back together.

Whatever form plate tectonics took back when the Acasta Gneiss was being baked into the Slave protocontinent, by the time the core of North America was forming around Superior, things were different. Probably, Archaean plate tectonics saw small, thin crustal plates moving and breaking up quite rapidly. In any case, by about 2.5 billion years ago, the plates had thickened, possibly by accretion of mantle material from below as Earth slowly cooled. Certainly, once Superior, Slave, Nain, Wyoming, Hearne, and Rae amalgamated, they stayed that way and the same holds broadly true for the ancient core fragments of South America, Africa, Baltica, Siberia, Antarctica, India, and Australia. They joined and broke apart many times, but the breaks and joins happened in much the same places. Each time, a bit more crust was added as island arcs and squished oceanic crust was smeared onto the edges, but the lines of weakness were still there. This explains why North America, for example, appears to grow by the addition of strips of ancient mountain chains plastered along the edges. The same things were happening over and over again, but at different times and in different places.

By 750 million years ago, Rodinia was complete. The mighty Grenville Mountains had risen to their dramatic heights and were now well on their way to becoming the eroded remnants we see today. Rodinia seemed settled, but not for long.

The first crack opened along what is now the west coast of North America. Australia, Antarctica, India, much of China, and parts of Africa decided to go it alone. A huge ocean, the Panthalassic (meaning all the seas), opened and North America (Laurentia), Siberia, Baltica, South America (Amazonia), and bits of Africa (Congo and West Africa) found themselves at the south pole. But the breakup wasn't finished. Around 570 million years ago, another crack appeared between Laurentia, Baltica, Amazonia, and Africa. The rifting followed old lines, running almost parallel to, but slightly east of, the ancient Grenville Mountains.

Iapetus grew as Laurentia pulled away from its neighbours, now forming their own supercontinent, Gondwana. By five hundred million years ago, Iapetus may have reached five thousand kilometres across. Scattered over it were volcanic islands and continental fragments. The fragments close to Laurentia included bits that would one day be called Northern Ireland and Scotland. Close to the Gondwana side lay what would become Southern Ireland and England, proving that political and cultural divisions go way back.

But Iapetus' days were numbered. Between 490 and 440 million years ago, four slivers of continent, Ganderia (part of South America), Avalonia and Meguma (parts of North Africa), split off from Gondwana and headed for Laurentia.

First some arcs of volcanoes and pieces of the Iapetus floor were smeared onto Newfoundland and the surrounding edges of Laurentia. Then Ganderia arrived between 450 and 430 million years ago. At the same time another microcontinent, Carolina, increased the size of what is now the southeast United States. Avalonia came in to join Ganderia around 421 million years ago and, finally, Meguma joined the party between 400 and 390 million years ago, completing the foundation of

Newfoundland and making it one of the most interesting places that Lithoprobe studied.

### Newfoundland Appalachians:
### Exposing the Opening and Closing of Ocean Basins

Newfoundland is not known as "the Rock" without cause. And the rocks that are well exposed there tell a complex and marvellous story of oceans opening, closing and then opening again. Lithoprobe images show the complexity throughout the crust.

Geologically, the Rock is divided into groups of rocks that have affinities with Laurentia and those that have affinities with Gondwana: the two major continental

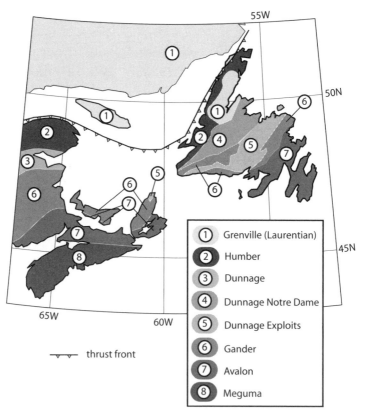

thrust front

① Grenville (Laurentian)

② Humber

③ Dunnage

④ Dunnage Notre Dame

⑤ Dunnage Exploits

⑥ Gander

⑦ Avalon

⑧ Meguma

Map of Newfoundland and maritime provinces showing the tectonic units that comprise the region. The circled "1" in western Newfoundland identifies the Long Range Inlier.

blocks, one in the north and one in the south, that eventually amalgamated about 270 million years ago to form the supercontinent Pangaea. Between the two blocks was the Iapetus Ocean, with oceanic crust lying underneath its waters, and peppered with volcanic arcs and islands much like the present southwest Pacific Ocean.

Sedimentary rocks that formed on the passive margin of Laurentia now form the Humber zone of western Newfoundland. That they were deposited on the Grenvillian crust of Laurentia is shown by the existence of a large chunk of Grenville rocks (the Long Range Inlier) exposed in the middle of the Humber zone as a result of faulting. Immediately to the southeast of the Humber lies the Notre Dame sub-zone of the Dunnage zone. Rocks in this region represent the volcanic arcs and pieces of oceanic crust of western Iapetus that were first smeared onto Laurentia.

As Ganderia arrived from the southeast, Iapetus was further closed such that its eastern arcs, islands, and ocean crust collided with Laurentia. The Exploits sub-zone of the Dunnage zone contains rocks representing the closure of the ocean and the building of a mountain belt as the microcontinent Ganderia collided with Laurentia. The Gander zone comprises the sedimentary rocks and crust of Ganderia. Not far behind another microcontinent, Avalonia (represented by the Avalon zone of eastern Newfoundland), continued the collision process, enlarging the mountain system. The last collision in this extended mountain-building process involved Meguma, a large tract of continental crust that has affinities with western Africa. In Canada, it is only found in the Maritime Provinces, not Newfoundland. Two hundred million years ago, rifting associated with the formation of the Atlantic Ocean initiated east of the Meguma zone, leaving Newfoundland and the Maritimes to tell the Appalachian story and leading to the present Atlantic passive margin.

A geological map of Newfoundland shows that it is composed of zones of rock: the Humber Zone is part of the Grenville Mountains that were altered by the pressure of the arrival of the microcontinents and Gondwana; the Dunnage Zone is the remnant of the volcanic islands and ocean floor; and the Gander and Avalon Zones are microcontinents. All these zones can be traced across the Gulf of St. Lawrence to the

**TIME 1**

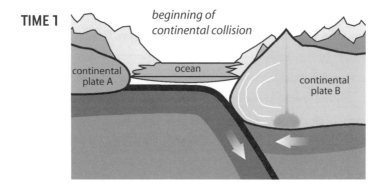

**TIME 2**

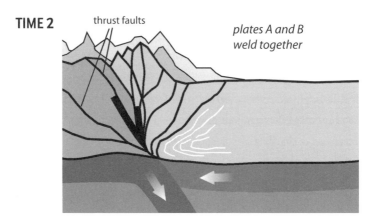

**TIME 3**

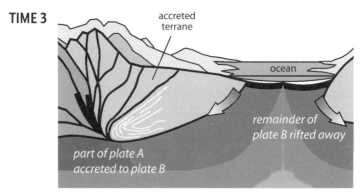

Schematic illustration of collision and rifting leading to an enlarged continent. A plate having both an oceanic and continental part, like the present North American plate, subducts beneath another continental plate (Time 1). The continental part cannot be subducted so there is a continental-continental collision, like the India–Eurasian collision, and the plates are welded together along a set of thrust faults (Time 2). At some later time, rifting and sea-floor spreading are initiated along a position different from that along which the collision took place, leaving a fragment of one continent accreted to the other, while the remainder is rifted away (Time 3). The Newfoundland and maritime Appalachians formed in a more complex version of this process.

Maritime provinces and New England states where they are joined by
the Meguma Zone. Thus, by running Lithoprobe lines from west to east
across Newfoundland and out to sea, it is possible to study the margin
of Laurentia as Iapetus opened (Humber Zone), the mountains formed
when Iapetus closed (Gander, Avalon, and Meguma Zones), and the
margin of North America as the Atlantic opens.

A minor piece of this story was the attachment of the oceanic crust
rocks, which Sawney Bean found so accommodating and I found so
interesting, to the edge of the Scottish fragment.

This mountain range, which was complete three hundred million
years ago when Laurentia and Gondwana had finally finished their
dance, once rivalled the Alps in size and scenic splendour. It is now
much reduced but still impressive in scale. The largest part runs from
Alabama to Newfoundland as the Appalachians, but other bits form the
Scottish Highlands, the Caledonian mountains of Scandinavia, and the
Atlas Mountains of North Africa. It was an immensely complex process,
taking place in many different stages as subduction zones formed,
microcontinents and island arcs crashed together, and sedimentary
basins formed, filled, and were destroyed. Because all this happened relatively recently, mere hundreds, rather than thousands, of millions of
years ago, we can see the complexity much better than in equivalent
earlier events such as the formation of the Grenville Mountains.

When Laurentia and Gondwana finally came together, they formed
the most recent supercontinent, Pangaea, which Alfred Wegener proposed in his 1915 book. It is easily the best-known supercontinent,
mainly because we are currently watching it break up. The Pangaean
rocks are still fresh and, consequently, geologists and geophysicists can
reconstruct events in the continent's history in more detail than any
other occurrence in the supercontinent cycle. What's more, it allows us
to speculate on the effect of the supercontinent cycle on the creatures
swimming or crawling around.

Pangaea stretched from pole to pole. If there is one supercontinent,
there must be one super ocean, another Panthalassic Ocean. Of course,

it's never as simple as that. Not all the continents joined Pangaea. To the east, bits and pieces that would one day form east Asia stayed away across the Tethys Sea. Nevertheless, a continent the size of Pangaea would have had a dramatic effect on climate, the environment, and the creatures doing their best to live on and around it.

The centre of Pangaea must have been very dry with extreme variations of temperature. Each end must have been cold, and several periods of glaciation occurred in its history. More subtly, collecting all the continents together greatly reduces the area of shallow seas around their edges. These seas, the modern continental shelves, are a vast storehouse of life and may be vital to its very development.

Life on Pangaea and in the Panthalassic Ocean and the Tethys Sea must have been tough. How tough can be seen in a 248-million-year-old apocalypse that wiped out 95 per cent of all life in the oceans along with 70 per cent of the creatures that had only recently ventured onto the land. It occurred as the Permian (290 to 248 million years ago) became the Triassic (248 to 206 million years ago), and it dwarfs the extinction that wiped out the dinosaurs. What went wrong?

The cause of the greatest threat to complex life on our planet since it blossomed almost six hundred million years ago is the subject of hot debate. Aficionados of the dramatic suggest  an extraterrestrial impact, although what evidence there is suggests that any impact around the right time was smaller than the one that killed the dinosaurs. Those with a more gradualist turn of mind point to evidence of glaciation, others to evidence of a greenhouse Earth. Perhaps the most likely culprits are the Siberian Traps. These are a vast outpouring of basaltic lavas that, if spread evenly over the world, would cover the surface to a depth of ten feet. The lavas must have been accompanied by an incredible belching of unpleasant gases that would have significantly altered the atmosphere and the chemistry of the oceans.

Whatever the cause, the effects were fundamental. The incredibly successful trilobites vanished, so much plant material died that coal formation ceased for millions of years, and life was thrown so far back on its heels that there was a brief age of fungus. As life will, it recovered, although in many strange and different forms and over a period of about ten million years.

If the Siberian Traps are to blame, it is very tempting to relate their incredible scale to the existence of Pangaea. We know heat likely builds up under supercontinents and that the protesting crust bulges and cracks open. It's easy enough to imagine lava pouring out of those cracks like pus from a boil.

Supercontinents are an extreme expression of plate tectonics. Undoubtedly, they cause extreme conditions on Earth's surface and have a profound effect on life. Maybe the great Permo-triassic extinction wasn't the only time the supercontinent cycle almost ended Earth's experiment with life.

Back when Rodinia's remnants hadn't finished deciding what they were going to do, another extinction came even closer to making Earth as barren as Mars, but, in doing so, it may have created a setting that made everything that came after, including us, possible.

Earth tolerates our existence thanks to an exquisite balance of forces. Tilt the world's axis a few degrees and the climate goes haywire; move an ocean current or two and the prairies become desert; alter the delicate equilibrium of the carbon dioxide cycle and we either freeze in a howling blizzard or bake in a humid greenhouse. Extraterrestrial encounters are fashionable these days and have affected life and its development, most notably ending the dinosaurs' reign. However, we mustn't lose sight of the less glamorous but equally effective causes of change.

We tend to think of the dramatic moments in evolution in terms of what failed to make it past the crises: the dinosaurs sixty-five million years ago or the trilobites 183 million years before that. The dinosaurs are to blame for this. Their remains have been known from long before we knew what they were or how they fitted into the grand scheme of things, and they are so impressive that their imagined exploits have been able to captivate generations of readers and movie-goers. How could such behemoths vanish with barely a trace?

Of course, there are many ways of looking at almost anything. If you see the glass as half full rather than half empty, these extinctions can be seen differently. The odd, lumbering amphibians and mammal-like reptiles of the Permian had to go to make way for the dinosaurs and the dinosaurs had to go to make way for the mammals and, ultimately, ourselves. So, if you are looking at things from the pinnacle of a successful spurt of evolution, mass extinctions are a good thing. Given that perspective, the most significant mass extinction in the history of life was neither the end of the dinosaurs nor that of the trilobites, but one far back in time, which destroyed ecosystems about which we have only a shadowy understanding. But first of all, let's get back to the *Titanic* (not the movie, the real one).

Shortly after 10 p.m. on a cold April night in 1912, a wireless operator, Jack Goodwin, on duty at the Marconi Company wireless station at Cape Race in Newfoundland, received a distress call. He listened intently then turned to a colleague: "My God, Gray," he said, "the *Titanic* has struck a berg!"[12]

Over the succeeding two and a half hours, the operators followed the drama being enacted four hundred miles west of them. They didn't yet know it, but as they listened and relayed messages, human folly, an iceberg, and the frigid waters of the North Atlantic were combining to end 1,513 lives. It might not have been the first time that ice had been a factor in ending life near Cape Race.

In the cliffs beneath where Jack Goodwin had his moment in the historical spotlight, and for the eight kilometres south to Mistaken Point, are a sequence of quite extraordinary rocks. They formed as mud deep beneath the growing Iapetus Ocean. Occasionally, volcanoes on nearby islands spewed out a layer of ash and the whole sequence was covered by clay that slumped down the sloping sea floor. What makes these rocks extraordinary is the incredibly rich and diverse collection of bizarre creatures that lived and died here.

Elkanah Billings is widely regarded as Canada's first paleontologist. Billings began his professional life as a lawyer, but soon gave it up to study fossils, of which he eventually catalogued 61 new genera and 1,065 new species. In 1855, Sir William Logan, who was considered the god of Canadian geology at the time, hired Billings as paleontologist for his young Geological Survey of Canada. But it was not all plain sailing. Perhaps Billings was not ideally suited to be a government employee. In April 27, 1869, Logan felt the need to send a memo to Billings: "Your constant absence from the office is a worrying annoyance, particularly as I have reason to suspect that it does not arrive from rheumatism."[13]

*Sir William Logan relaxes with some ornate furniture after having put Canadian geology on the map.*

William Notman/Library and Archives Canada/C-010418

### The Geological Survey: Canada's Oldest Scientific Organization

Canadian Earth Science is recognized internationally for the quality and quantity of its research, and has been since the mid-1800s when the Geological Survey of Canada was founded. The survey is the primary government-based Earth

*Canadian exhibit at London World Fair 1851*

Science organization in Canada and dates back to a time before Upper and Lower Canada and the Atlantic provinces joined to form a Dominion—in fact, back to September 1841. At that time, the Legislature of the Province of Canada, which had formed only a year earlier by amalgamation of Upper and Lower Canada, resolved "that a sum not exceeding one thousand five hundred pounds sterling be granted to Her Majesty to defray the probable expense in causing a Geological Survey of the Province to be made." Even then, the Legislature was trying to catch up with the United States and Britain, each of which had undertaken surveys in the early 1800s that established mineral resources and new agricultural lands. So the organization was founded in the following year, 1842, with William E. Logan being appointed to carry out the survey of the Province. Logan was tremendously successful. He is recognized as Canada's first great scientist. For the 1851 London World Fair, he put together an impressive display of Canada's minerals. He followed this with an even more impressive display at the 1855 Exhibition in Paris, which included the first geological map of Canada, and so impressed Queen Victoria that she knighted him—Canada's original "rock star." By the time of his retirement in 1869, Logan had firmly established both the organization and the reputation of the Geological Survey of Canada. Mount Logan in the Yukon, the highest peak in Canada, is named after William Logan.

---

This slap on the wrist doesn't appear to have bothered Billings, who went on collecting and cataloguing his beloved fossils. In 1872, on Duckworth Street in downtown St. John's, Billings chipped an odd, disc-like structure out of the rock. In honour of where he found it, he called it *Aspidella terranovica*. It was frequently dismissed as a pseudo-fossil, a structure in a rock that mimics a fossil but which has an inorganic origin, but as more and more were found around the world, people began to take notice.

Billings had stumbled on the first Ediacaran fossil. Named for a rich locality in Australia, these odd beasts have turned up all over the world, on every continent except Antarctica. Although they may have extended both backwards and forwards in time, they thrived between 565 and 543 million years ago, immediately before the Cambrian (543 to 490 million years ago) gave us the apparent explosion of life with hard parts that can be preserved relatively easily in rocks. Nowhere are the Ediacarans better seen than in the area around Cape Race.

The Ediacarans were a weird bunch, ranging from discs, to fern- or frond-like shapes, and worms. They looked like plants but were animals, living out their lives attached to the ocean floor, filtering what nutrients they could from the passing waters. Most remarkable is the fact that we know anything about them at all.

The Ediacarans had no shells or hard body parts. It is reasonable enough to imagine a dinosaur bone sinking to the bed of a river, being covered by sediment and preserved over the millennia. It's tougher to imagine that happening to a jellyfish. It is possible, but fossils of soft-bodied animals are extremely rare, yet here, preserved worldwide, is a complete ecosystem of creatures you could squish in your hand.

On top of that, it's hard to say what the Ediacarans are related to. Some look superficially like animals that came later. Most, however, are so different that some researchers have even suggested that they represent neither plants nor animals, but a completely separate kingdom that no longer exists.

"What are the Ediacarans, and why are so many of them preserved?" are perfectly valid questions but perhaps more intriguing is, "How did they come about at all?" For most of its history, Earth has been populated by simple life, single cells, or strands of algae. Suddenly, sometime shortly after six hundred million years ago, complexly organized, multicellular organisms, the Ediacarans, proliferate. The body plans of almost all living animals originated suddenly in an unbelievable burst of evolutionary activity between about 600 and 525 million years ago. Why? Or at least, why then and not five hundred million years earlier or later?

Obviously, the chemistries of the atmosphere and ocean were suitable, and the moving continents provided friendly environments. The Ediacarans apparently had no predators, but such a dramatic, global event must have had a global cause. Perhaps it was something even more controversial than Billings' discovery of such a strange form of life in the first place.

In Canada we know a lot about snow, but it comes and goes. Admittedly, many places have too much of it for too long, but even the bleakest places in January look immeasurably better in August. This is just as well because if snow doesn't melt in the summer, we're in trouble.

Earth receives about 343 watts per square metre in radiation from the Sun. Some never makes it through the clouds and some is reflected back from the surface, but about two-thirds is absorbed to keep us warm. Of course, there is a balance and Earth emits radiation to stop us frying, but not so much that we freeze. Upset that balance and strange things happen.

Light-coloured objects reflect radiation better than dark ones. Therefore snow reflects more solar radiation than a tree-covered hillside. Cover that hillside with snow, and you reflect more radiation. Cover that hillside and a whole bunch of other hillsides with snow all year round and the amount of reflected radiation is enough to cool the climate. Cool the climate and you get more snow. More snow, more white hillsides, more reflected radiation, cooler climate. The process feeds on itself until a balance is found. This is what happens during an Ice Age. But there is a theoretical point beyond which the process becomes irreversible and we are in serious trouble.

If continental glaciers and sea ice extend from the poles to latitude 30 north and south, then half Earth's surface will be white and very reflective. After that the process is unstoppable and the entire globe will end up covered with hundreds of metres of ice and snow.

The idea that it was possible to turn Earth into a giant snowball was first suggested in the 1960s by Brian Harland of Cambridge University. Obviously, it had never happened, for two reasons: a snowball Earth would kill off all life, and microscopic fossil life goes back 3.5 billion years. With a completely white, highly reflective Earth, how could the process ever reverse itself? Or was it that simple?

In 1977, the US Navy–owned research submersible *Alvin* was being steered around on the ocean floor near the Galapagos Islands, investigating warm water plumes. *Alvin*'s operators discovered that the source of the plumes were black smokers: vents of superheated, sulphide-rich water pumping out of vents associated with the spreading ridges between oceanic plates. Black smokers from depths of thousands of metres reach temperatures of hundreds of degrees Celsius, and the precipitating sulphides form strange shapes equivalent in height to a thirteen-story building. They are remarkable things, not least because of the wonderful array of life that thrives in their inhospitable shadow. Bacteria, tube worms, clams, mussels, crabs and blind shrimp happily go about their business in this warm, but totally black, environment. The significant thing from our point of view is that these creatures derive all the energy they need for life from the vent, not from the Sun. Therefore, they wouldn't care if the sea far above were encased in ice. A snowball Earth would mean little to these beasts as long as their vent home kept pumping out its hot soup.

The black smokers exist because of plate tectonics. So do volcanoes, and they would go on spewing out lava and large amounts of carbon dioxide whether the surface of Earth was frozen or not. Since a frozen Earth would have no free water on the surface to absorb the carbon dioxide, it would have nowhere to go and build up to incredibly high levels. It would take millions of years, but the result would be a ferocious greenhouse effect that would raise the temperature enough to melt the ice.

So a snowball Earth would not kill all life on the planet and would not last forever. Earth's surface would flip-flop from extreme cold to extreme heat. But has it ever happened?

Many of Pangaea's rocks show evidence of having been glaciated. According to plate reconstructions, some of these rocks were close to the equator when covered with ice, suggesting glaciation far more widespread than recent ice ages. There are also some interesting rocks in Namibia.

Namibia hosts some well-developed glacial sediments from pre-Ediacara times. The odd thing about them is that they are immediately capped by a carbonate rock called dolostone. The dolostone formed very quickly in warm water and is rich in inorganic carbon. All those factors fit perfectly with a sudden flip from a cold ice age to a hot greenhouse world rich in carbon dioxide.

So, maybe the world *has* been a huge snowball in the past—perhaps several times. What does that mean?

Despite the carnage wrought by a snowball Earth, life survived—bacteria lived on at deep hydrothermal vents, photosynthetic eukaryotic algae clung to life around hot springs scattered across volcanically active areas, and cold-loving organisms, like the ones that today thrive in Antarctic valleys and around dust particles in floating ice, didn't even notice what had happened. Then the ice melted. The world was suddenly warm, nutrient-rich, and inhabited by a tiny population of very hardy, well-adapted life forms. They went crazy, proliferating wildly and spreading everywhere. Evolution tried out a wealth of different shapes and sizes, some of which failed or got eaten when the predator experiment began, but some survived to continue a more sedate evolution that led to a creature that can look back and speculate on all this.

The uniformitarians wouldn't like it, but it might have happened. If it did, the snowball Earth and its dramatic evolutionary consequences were a product of plate tectonics, of the creation and destruction of Pangaea and Iapetus. The story that Lithoprobe unravelled illuminates plate tectonics, which in turn enhances our understanding of life's tortuous journey and our own beginnings. But does it have any immediate, modern-day relevance? Of course it does, at least to the small patch of North America beneath my feet as I write this.

# A Wilderness Tale
by John Percival

During the summer of 1985 I planned an expedition to the Fraserdale-Moosonee block, at the northern end of the Kapuskasing uplift in northern Ontario, to map the area and collect samples for study. This poorly exposed and inaccessible region is traversed by the North French River, where bedrock is exposed at rapids and waterfalls. My assistant and I flew by float plane into a lake at the south end of the river, and selected a small lake just off the river for pickup five days later. Due to its remoteness, there is no canoe route map, so even the modern voyageur has to rely on old tree markings to identify upcoming hazards and portages. On the first morning I had maps, air photos, and a notebook on the seat beside me for easy access as we paddled gently downstream, searching for outcrops. Focusing thus on geological features, I missed an important navigational prompt and we found ourselves in fast-moving water that soon turned into rapids, which we managed to shoot without difficulty. What turned out to be more troublesome, however, was the small waterfall at the foot of the rapids.

Although we tried valiantly to propel the canoe safely over the obstacle, it sank momentarily... long enough to liberate all navigational materials and soak food supplies. Fortunately, the day was sunny and warm, and we were able to dry out clothing and bread within a few hours (the effect on the cookies was more disastrous). The immediate problem, however, once the paddles were recovered, was navigation: five days to go, and no maps. We weren't worried about getting lost, just how to do geology? Using river azimuth, landmarks (rapids, waterfalls) and approximate distances, and with the help of air photos back in the office, I was later able to locate geological observations and sample localities. There were

numerous other water incidents on this wild river, but none as significant as losing all the maps.

On the appointed pickup day we reached a straight, shallow stretch
of the river, which I recalled as being adjacent to the intended lake, and
we watched intently for the connecting creek. Panic began to set in as the
pick-up time approached and we searched in vain for the subtle land-
mark. At this point I have to believe that divine intervention played a
role in the outcome of the expedition. We could not believe our eyes
when a motor boat approached from downstream (the river is navigable
to Moosonee, several days away by canoe). Two native hunters were on
board and knew the territory well. Their buddy had a trap line on the
lake we were looking for and they knew the location of the portage (the
creek was a trickle). We thanked them profusely and made it to the
pickup spot with ten minutes to spare.

## CHAPTER 7

# OUR VERY OWN MOUNTAINS

The great Thunderbird finally carried the weighty animal to
its nest in the lofty mountains and there was the final and terrible
contest fought. There was shaking, jumping up and trembling of
Earth beneath, and a rolling of the great waters.

—Northwest Washington

First Nations Myth

They had practically no way or time to try and save themselves. I
think it was at nighttime that the land shook.... I think a big wave
smashed into the beach. The Pachena Bay people were lost ... But
they who lived at Ma:lts'a:s, House-Up-Against-Hill, the wave did not
reach because they were on high ground ... Because of that they
came out alive. They did not drift out to sea with the others.

—Vancouver Island First Nations historical account of the destruction of
the community of Pachena Bay as related by Louis Clamhouse, 1964.

TALES OF THE EPIC STRUGGLE between the Thunderbird and Whale
are common all along the Pacific Northwest coast. Sometimes the Whale
dives, taking the Thunderbird to bottom of the ocean. Sometimes the

Thunderbird lifts the Whale to his nest and defeats him there. Either way, these supernatural beings cause shaking of Earth, landslides, and huge waves, all of which can result in terrible losses among the puny humans caught up in them.

The Thunderbird and Whale are unique to the western coast of North America, but similar myths are told in earthquake zones worldwide. In Sumatra there is a struggle between a horned monster and a giant serpent; in Japan it is a huge catfish and the god Kashima. Even if there is no battle, there are still mythical explanations for the unstable ground. The Hindus believe it is one of the eight mighty elephants that hold up the world shaking its head, some Siberians say that it is the god Tuli's dogs scratching fleas, and in Mongolia it is a twitching giant frog. But the ancient Greeks came up with the most elaborate theory: that earthquakes were caused by the movement of air through immense subterranean chambers.

Our myth is plate tectonics and it is certainly closer to a scientific truth than the others. Nevertheless, the oral myths and legends of First Nations peoples contain some fascinating hints and clues about a time before there were any scientists to take notes.

As we have seen, subduction is a vital element of plate tectonics and may even have been crucial to the survival and development of life on Earth. The downside is that it can be deadly.

On holiday once in Rotorua in New Zealand, I had just settled into my motel room and was looking forward to visiting the famous hot springs when the people in the room above began jumping on the floor. My first reaction was to hope that whoever was up there didn't plan on doing the same at two in the morning, but then I realized with a chill that I was in a single story motel. With all the horror movies I had ever seen replaying in my mind, I went out, checked that there was indeed no upper story and went to the office. "Oh, it was just an earthquake," the desk clerk informed me. "There have been a lot recently, ever since the new geyser broke through the street a block away."

The Rotorua earthquake was caused by adjustments in hot water under pressure, and was very minor. I have experienced two stronger earthquakes on Vancouver Island. In both cases, I was much farther from the epicentre and they felt like a car hitting the side of my house and the windows rattling in a strong wind. But my father experienced something entirely different.

On the afternoon of January 15, 1934, the Indian Plate jerked against Asia. The Great Bihar earthquake measured 8.1 on the Richter scale. It killed tens of thousands of people and devastated towns across north India and Nepal. My mother saw the chandelier of her house sway, knew what it meant, grabbed my infant sister and fled onto the front lawn moments before the bungalow crashed to the ground. My father watched cracks open in the ground and saw railway lines and bridges—his particular concern as a civil engineer—destroyed. I have sepia photographs of railway lines snaking across a plain as if they were string flicked by a playful child.

Two years later, my father was home on leave in Scotland and went to see a movie at the Odeon cinema on Renfield Street in Glasgow. A few minutes into the movie, he heard an earthquake coming. Reacting as he had in India, he leapt out of his seat and sprinted up the aisle, only to realize that the rest of the audience was staring at him openmouthed. As he hesitated, the subway train from nearby Queen Street station rumbled beneath the cinema. Sheepishly, he returned to his seat.

Earthquakes can result in anything from holes in the road in New Zealand to the complete destruction of cities. They can cause embarrassment in movie theatres or cultural trauma lasting generations. Any two pieces of rock moving against each other can precipitate an earthquake, but there is one class of earth movement that is different from all the others. The most violent earthquakes ever recorded occur where warm, buoyant, and relatively new crust is being forced rapidly down into a subduction zone. These are megathrust earthquakes and there have been five in the past century:

- 1952 Kamchatka Earthquake, magnitude 9.0—Pacific Plate subducting beneath the Okhotsk Plate.

- 1957 Andreanof Islands Earthquake (Aleutian Islands), magnitude 9.1—Pacific Plate subducting beneath the North American Plate.
- 1960 Great Chilean Earthquake, magnitude 9.5—Nazca Plate subducting beneath the South American Plate.
- 1964 Good Friday Earthquake (Alaska), magnitude 9.2—Pacific Plate subducting beneath the North American Plate.
- 2004 Sumatra-Andaman Earthquake, magnitude 9.15—Indo-Australian Plate subducting beneath the Burma Plate.

Each of these events is immense on a human scale. Each triggered a tsunami that caused widespread death and destruction. But only the most recent one has occurred when sophisticated systems were in place to measure the full impact of one of these events.

One hundred and sixty kilometres west of Sumatra and thirty kilometres beneath the surface of the Indian Ocean, the vast Indo-Australian Plate is being forced beneath the Burma Plate. Until recently, the plates had been welded together for many years, but the pressure was building. Eventually, at one minute and seven seconds before eight in the morning, local time, on Boxing Day 2004, it all became too much. Perhaps the slip was triggered by a magnitude 8.1 earthquake near New Zealand on the opposite side of the Indo-Australian Plate, three days before, or perhaps it was just time. In any case, the weld ruptured.

The rupture began just north of Simeulue Island and sped north at up to one hundred thousand kilometres per hour. After one minute and forty seconds, as if the Earth were drawing breath, everything stopped. Another minute and a half later, the rupture set off again, slower this time, only about 7,500 kilometres per hour, but travelling much farther.

In total, during the almost ten minutes of the Sumatra-Andaman earthquake, Earth's crust ruptured over 1,200 kilometres, moving horizontally up to ten metres and vertically by four to five metres. The quake released enough energy to power the entire United States for eleven days, it moved Earth's surface three millimetres vertically in Oklahoma, wobbled Earth's axis by 2.5 centimetres, altered Earth's rotational speed enough to shorten the day by 2.68 microseconds, created one-and-a-half-kilometre-high thrust ridges on the Indian Ocean floor, dug a new oceanic trench, and moved the northern tip of Sumatra by as much as thirty-six metres to the southwest. But all this says nothing about the human cost.

The vertical movement on the sea floor displaced some thirty cubic kilometres of sea water that sped away from the fault line in a series of waves. In deep water, the waves reached heights of sixty centimetres and travelled at speeds of up to 750 kilometres per hour. In shallower water, close to land, the waves slowed to a few tens of kilometres per hour, but built up to heights of thirty metres in places.

---

### Tsunamis: Killer Sea Waves

A tsunami (pronounced *sooh-nahm-ee*, from the Japanese word for "harbour wave") is a sea wave generated by major under-sea earthquakes, monstrous volcanic eruptions or huge submarine landslides. However, most major tsunamis are caused by large subduction earthquakes of magnitude eight or greater on the Richter scale, which can offset the sea floor by a few metres or more.

Above the displaced sea floor, all of the ocean is affected so the entire body of water is involved in the wave motion, not just the surface as is the case with wind waves. In the open ocean, the tsunami wave is less than one to two metres from crest to trough, just a slight undulation. It also has very long wavelengths (the distance between two successive peaks), as much as one thousand kilometres, so a ship at sea would never notice it. But it travels amazingly fast—a speed of five hundred to eight hundred kilometres per hour in the open ocean, depending upon water depth; speeds that are equivalent to the fastest commercial jet planes. Another important aspect of a tsunami is that it is made up of a series of concentric waves, not just one wave, rather like the ripples you see when you throw a rock into a pond.

Eventually the tsunami will reach a region where the depth of water decreases. As the water becomes shallower, the wavelengths become shorter, the speed slows down but the *height* of the wave increases. This effect is most dramatic and hugely damaging as the tsunami waves reach shallow water near the shoreline. There, the waves slow to tens of kilometres per hour—not jet plane speed any more, but faster than you can run, and the wavelengths reduce to tens of kilometres. But at the shoreline, the height of the waves from crest to trough can increase up to a few tens of metres. The waves become a wall of fast-moving water.

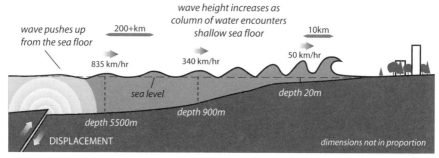

Schematic illustration of the development of a tsunami generated by a large sea-floor earthquake. Wave heights may be less than a metre in the open ocean. Along a shoreline, the wave heights could be thirty metres, a veritable wall of water.

Tsunamis can devastate areas so far from their source that no one feels the earthquake. Often the first sign of a tsunami is not the gigantic sea wave but the sudden withdrawal of water from the shore, far below the normal low tide variation. This is the trough part of the wave and precedes the wall of water that will arrive a few minutes later. The crest of the powerful wave will surge over everything in its path and can proceed inland for a kilometre or more, depending upon the near-shore topography. Remember the concentric waves? The wall of water rapidly recedes, dragging everything in it out to sea; this is another trough. But another crest can follow, destroying whatever wasn't hit the first time. This can happen a few times, with diminishing size and energy, but by this time the damage has already been done.

DigitalGlobe/Getty Images

*The consequences of a magnitude 9.15 megathrust earthquake. Banda Aceh before and after the Sumatra-Andaman Earthquake of Boxing Day 2004.*

The Boxing Day tsunami was the most devastating in recent history. It came ashore with a combined force greater than the entire explosive energy used during all of World War II (including the two atom bombs), killing 283,100 people, a disproportionate number of whom were women and children. It contaminated drinking water, destroyed fields, significantly affected the economies of several countries, and displaced over a million people. The psychological trauma of the disaster caused an upsurge in ghost sightings, has been cited as evidence of God's displeasure with the Asian sex trade, and, according to Egyptian weekly paper, *Al-Osboa'*, might be the beginning of "experiments to exterminate humankind" by India with the support of the United States and Israel.

Obviously, the Sumatra-Andaman earthquake was one of the greatest natural disasters in recorded history, but it is not exceptional. It has happened before and it will happen again. And next time, it might be much closer to home.

Perhaps 95 per cent of the First Nations oral history that was extant when first contact was made along the Pacific Northwest coast has been lost. That is an immense tragedy, both from a cultural and a historical perspective, but the remnants can still tell us something of value. About 120 surviving stories from northern California to Vancouver Island refer in some way to the ground shaking violently or flooding that is consistent with a tsunami. The most detailed of these can be subdivided into three categories: historical tales with enough information to suggest a date (e.g., in the time of a great-grandparent etc.); historical tales that cannot be dated; and myths, like that of the Thunderbird and Whale, that have absorbed fragments of fact.

The historical tales are vivid and dramatic, describing entire villages washed away or buried beneath landslides, people fleeing to high ground or in canoes and being unable to stand upright because of the shaking. Interestingly, these stories vary up and down the coast. In California, many stories talk both of earthquakes and tsunamis, whereas on Vancouver Island these tales are separate. Between, earthquake stories are rare, while flood stories are common. In part this probably reflects cultural differences and preservational factors, but it may also be the result of a variation of earthquake effects along the coast. For example, if the earthquake occurred close to shore off California, the Earth shaking and tsunami would happen very close together in time, naturally seeming to be one event. On the other hand, if the earthquake off Vancouver Island occurred farther offshore, there would be more of a time lag between the shaking and the flooding, encouraging separate stories.

Dating the events in these tales is a chancy business and individual dates for the most recent story range from 1400 to 1825. However, the average date is 1690 and can be refined to 1701 if the two most extreme estimates are removed. Is this an oral record, preserved over three hundred years, of an event so vast and frightening that it was indelibly imprinted on the cultural memory of those who survived? Are there ways to confirm this and tie down the date more precisely?

Several hundred kilometres offshore from the Pacific Northwest, the Juan de Fuca Ridge rises almost three thousand metres from the sea floor to form a stepped chain of undersea mountains. Like the ridge in the centre of the Atlantic, this is a spreading centre, with the Pacific Plate being pushed to the west and the Juan de Fuca Plate to the east at five to six centimetres per year. Unfortunately, there is not a lot of room between the ridge and the approaching mass of North America, so the Juan de Fuca Plate is being destroyed. Around eighty kilometres offshore from Vancouver Island, Juan de Fuca dives down beneath North America in just the same way as the Indo-Australian Plate dives beneath the Burma Plate off the coast of Sumatra. In just the same way, there is the potential for megathrust earthquakes.

Earthquakes with the power to wobble the entire Earth leave a mark. Even if all the sophisticated measurements that scientists made of the Sumatra-Andaman earthquake are lost, evidence will remain. Probably already, people in remote areas of Sumatra are telling stories about the time the Earth shook and the sea raced inland and trying to fit these events with their cultural beliefs. Even after Banda Aceh and the Thai tourist resorts have been completely rebuilt, a testament to the horrific destructive power of the earthquake will be preserved in the geological and environmental record for future researchers to wonder at. With luck, future researchers will have a time scale with which to work.

First Nations oral histories remember events other than earthquakes. The Klamath First Nations in Oregon tell a story about how the Chief of the Below World, who lived beneath a vast mountain, once fell in love with a beautiful human girl. She spurned him, and in a rage the Chief of the Below World vowed to destroy the Klamath people with fire. The Chief of the Above World took pity on the humans and gave battle on their behalf. Eventually, as mighty explosions echoed across the landscape and fiery rocks the size of dwellings flew through the air, the Chief of the Above World triumphed, forcing his opponent back beneath the ground and collapsing the mountain on him, trapping him forever. The mountain was called Mazama and where it collapsed sits Crater Lake.

Mount Mazama exploded 7,770 years ago, spewing ash over more than a million and a quarter square kilometres of Oregon, Washington, California, Idaho, Montana, Nevada, Wyoming, British Columbia, Alberta, Saskatchewan, and reaching as far east as Nebraska. While this must have been an appalling experience for the Klamath, the effects are a boon to scientists. Throughout the Pacific Northwest, there is a distinctive and readily datable marker, preserved amidst whatever was going on locally. Mazama ash is invaluable in dating deposits as varied as ancient lakeside cave deposits in Oregon and bison kill sites in Alberta. Mazama ash is also found in a thin layer buried in the deep sea mud and sands off the coast of Oregon and Washington.

Large rivers, such as the Columbia, carry huge amounts of sediment that are swept out to sea onto the continental shelf. The sediment builds up until, eventually, it becomes unstable and slips off the shelf into deeper water. The slump is called a turbidite and it leaves a very distinctive deposit on the ocean floor. When a number of turbidites are triggered together, for example by a large earthquake, they form a widespread blanket of deposit. Those turbidite blankets are separated by layers of fine mud in sediment cores of the ocean bed. Map the blankets and you have a record of these big turbidite events and of the earthquakes that caused them.

Since the ash from Mount Mazama drifted down through the water almost eight thousand years ago, there have been thirteen major turbidite events off the Pacific Northwest coast. That's thirteen major earthquakes, or one every 597 years, on average. Of course, averages don't mean a lot when dealing with specific events, and radiocarbon dating gives intervals between the last five earthquakes of 610, 980, 500 and 390 years from the oldest to youngest event. The most recent event occurred about 300 years ago.

### The Cascadia Megathrust Earthquake Cycle

The January 26, 1700, subduction or megathrust earthquake ravaged the west coast of Canada and the U.S. Pacific Northwest. Its estimated magnitude was greater than 9 on the Richter scale. It probably lifted the sea floor by metres and ruptured it for hundreds of kilometres offshore along the coast, generating a tsunami that was observed as far away as Japan. No such earthquake has occurred since, but the next "big one" will come. Unfortunately, we simply do not know when.

What causes these earthquakes and why are they cyclical? First we look to the west, to the Pacific Ocean. More than two hundred kilometres offshore, the Juan de Fuca ridge is an oceanic spreading centre, generating part of the Pacific plate, which is moving to the west, and the Juan de Fuca plate, which is moving to the east. Vancouver Island lies on the western part of the North American plate, which is moving westward. The Juan de Fuca and North American plates form an oceanic-continental collision zone. The oceanic plate is being forced below the overriding North American plate, forming a subduction zone. The boundary between these two plates is the surface along which the megathrust earthquakes occur.

At present, and throughout most of those 7,770 years, the subduction boundary between the two plates is locked; they are not slipping past each other. Yet the Juan de Fuca and North American plates continue their unstoppable eastward and westward drifting, respectively, at a convergent rate of

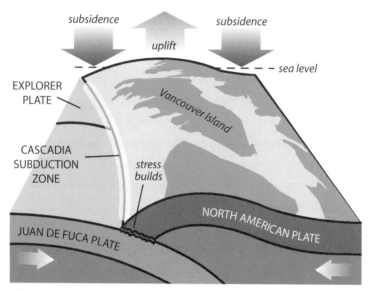

**Plate surfaces locked**

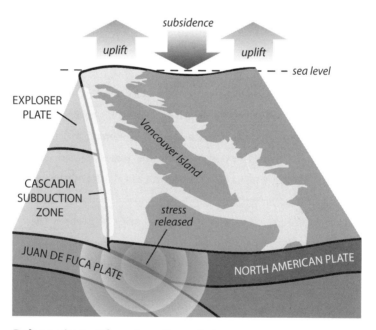

**Release (megathrust earthquake)**

Schematic illustration of the megathrust earthquake cycle off Canada's west coast. Stress builds up over a long period of time when the plates are locked, causing uplift and subsidence as shown. This stress is released instantaneously when the earthquake occurs, causing a rapid reversal of the uplift and subsidence from before. Following the earthquake, the plates are once more locked and the stress again builds up until another earthquake is generated.

about forty millimetres per year. While this rate doesn't sound like much, it amounts to almost twenty-four metres during the average 597-year period between megathrust earthquakes. The convergence causes stress and deformation of the two blocks. As a result, Vancouver Island is uplifting slightly, about one to four millimetres per year. To its west and east, there is a very small amount of subsidence. Laterally, Vancouver Island and its offshore margin are getting closer to the mainland by a few centimetres per year. This strain puts incredible stresses upon the rocks and the locked frictional boundary between the plates. Generally, rocks are strong and can elastically store the strain. However, at some point, the stresses become too much, the locked boundary, which is located below the continental shelf and slope, breaks instantaneously and a megathrust earthquake is generated. As the stress is relieved, the North America plate "snaps" upward to a low stress state. This causes immediate subsidence of Vancouver Island with an elastic rebound to the west and east. The abrupt rise of the sea floor due to the upward "snap" at the offshore fault generates a tsunami. The former bulge that is centred near the coast suddenly collapses, allowing the inflow of salt water. Over time, the subsided coastal area is filled with sediment. The interminable forces of convergence continue and the cycle of tectonic stress buildup and release starts over again.

How do we know the megathrust earthquakes are cyclical? The answer follows from some excellent Earth Science detective work. "Ghost forests," stands of the skeletons of dead cedars, were found to exist in tidal marshes. How did they grow and what killed them? In the late 1980s, Dr. Brian Atwater of the U.S. Geological Survey provided an explanation. He found that the dead trees were not rooted in the modern marsh. Their roots were founded in an older marsh that was buried by up to two metres of tidal mud. He found that the surface of the older marsh was marked by a thin layer of peat ubiquitous in the region of the dead cedars. Subsequently, he and other scientists in Canada and the U.S. found the same peat layer at numerous locations from central Vancouver Island to northern California.

What the scientists were observing was the result of a megathrust earthquake and its associated tsunami. For hundreds of years, the marshland surface near the shore in bays was forming, just as we would see it today. Then the earthquake occurs, causing the subsidence described above. A tsunami carrying lots of sand within its waves deposits the sand on the subsided marsh surface. Over time, the marsh deposit is changed to peat while another marsh surface forms above the sand layer. Many such peat and sand layers are found. Carbon dating allows us to determine the approximate age of the peat.

The ghost forest is caused in a similar fashion. Coastal cedars grow for hundreds of years, becoming tall and thick. A megathrust earthquake occurs, causing subsidence of the land on which they are growing. Salt water flows into the land, killing the trees. After hundreds of years, another marsh surface forms as the land slowly uplifts once again. All that is left of the once-magnificent cedars is their grey skeletons poking up out of the marshlands.

The dead cedars also function as clocks. Dendrochronology is the science of dating by tree rings. Comparing the pattern of tree rings in the dead trees to those of living trees that were equivalently old, the scientists determined that

January 1700 —

700 CE—

300 CE—

600 BCE —

1000 BCE —

*Photo from the book, The Orphan Tsunami of 1700, with permission from the author, Brian Atwater of the U.S. Geological Survey.*

*Evidence for five large earthquakes can be seen in this exposed tidal channel in Washington state. Fossil tidal marsh soils (dark horizons, labelled) represent old marsh surfaces that subsided about one to two metres during the earthquakes (dates indicated). Tidal mud (light-coloured areas) buried the marsh surfaces after the earthquakes. The top buried marsh soil records the great subduction earthquake of January 26, 1700.*

the last growing season for the dead forest was 1699. Thus, the trees likely died in the winter of 1699–1700.

The hypothesis of irregular but cyclical megathrust earthquakes also received support from offshore studies on the deep ocean floor. There, samples of sea floor sediments were acquired in long tubes by scientists from the University of Oregon. As expected, the cores contained fine-grained mud, which results from the slow, continuous settling of fine sediment in the ocean above. But unexpectedly, the scientists found unusual deposits of sand layers. The same patterns were found in sediment cores from a very large area. Dr. John Adams of the Geological Survey of Canada provided an explanation. He suggested that the sand layers were turbidites formed when sediments from the continental shelf were carried down the continental slope into the deep ocean by huge sub-marine landslides caused by very large earthquakes. Thirteen such layers were found above the well-dated Mazama ash layer, enabling the average interval between megathrust earthquakes to be estimated.

— Tidal Marsh

— Tsunami sand layer

— Fossil tidal marsh

*Detail of the bed of sand left by the tsunami from the January 1700 earthquake. The sand is exposed in a shallow pit dug at a tidal marsh near Tofino on the west coast of Vancouver Island. It was deposited on an older marsh that subsided during the earthquake and then was covered by tidal mud. The present tidal marsh layer is at the top.*

When a subduction zone is locked, the edges of the plates bulge upward. When a megathrust earthquake releases the pressure, the land subsides. When coastal plains subside they are flooded by sea water, which kills the vegetation, forming a layer of peat. If there is a tsunami, the peat is buried by a layer of sand and the process is repeated. On the Washington and Oregon coast there is evidence of seven cycles of sudden subsidence over the last 3,500 years with an average interval of five hundred years. Some of the tsunami sand sheets suggest waves ten metres high and are very similar to what occurred on the coast of Chile after the megathrust earthquake there in 1960. The most recent peat/tsunami sand layer in the Pacific Northwest has been dated to

*A ghost forest—a stand of weather-beaten dead tree trunks found in the marshes of southern Washington state.*

Photo from the book, *The Orphan Tsunami of 1700*, with permission from the author, Brian Atwater of the U.S. Geological Survey.

between 310 to 420 years ago and preserves the remains of First Nations campfires below it.

The trees that are killed when salt water floods coastal lowlands form ghost forests that can be dated. Radiocarbon dating of nine Sitka spruce trees from two different locations on the Washington coast show that all the trees died in a single event between 1680 and 1720.

To refine the date even further, tree rings were studied from living Western red cedars dating back to 993. These were compared to seventy-five dead trees from the ghost forest. All the dead trees died after 1680 and, where the outermost rings were preserved beneath bark, they were found to have died between the growing seasons of 1699 and 1700. So, we have narrowed down the last of a series of megathrust earthquakes on the Pacific Northwest to a specific winter. And even more is possible.

On a calm, sunny day in the twelfth year of the Genroku era, during the affection month, Mutsuki, a village elder in Miho, 145 kilometres south of Tokyo, wrote about a particularly high tide in his chronicle of important events.

"The water also went into the pine trees of Ego. The receding water went out very fast, like a big river. It came in about seven times before 10 a.m. of that day and gradually lost its power . . . Because the way the tide came in was so unusual, and was in fact unheard of, I advised the villagers to escape to Miho Shrine . . . It is said that when an earthquake happens, something like large swells result, but there was no earthquake in either the village or nearby."[14]

Other chroniclers record similar events, houses damaged, rice paddies engulfed, and storehouses flooded, all around the same time. Taken together, historical records along Japan's Pacific coast suggest a wave between two and three metres in height, unrelated to any earthquake or storm. To create a wave of such uniform height over such a large area, the source must have been very powerful and very far away: a megathrust earthquake somewhere on the Pacific rim.

Chile, Kamchatka and Alaska are all locations of historical megathrust earthquakes, but none produced this tsunami pattern in Japan. In addition, no Spanish records in Chile, Russian records in Kamchatka or oral histo-ries among Alaskan First Nations make any mention of an earthquake close to the right time. The only possible location is the Pacific Northwest.

If the Japanese date is converted to the western calendar, it comes out as January 27, 1700—right in the middle of the season predicted by the ghost-forest tree rings in Washington. If tsunami travel time is worked backwards, we have around 9:30 p.m. on January 26, 1700. This was the night that the Pachena Bay people died.

The Pacific Northwest is not a place where geology and plate tec-tonics can be taken for granted. It is a place where the lives of thousands of people depend upon our knowledge of what is going on beneath our feet. And, of course, Lithoprobe had a look. Actually, Lithoprobe scien-tists did much more than look at where the next big earthquake will occur, they took a long hard look at what is happening beneath a moun-tain chain that is still growing and how the west coast of North America formed. They did this across southern British Columbia where, on the west coast, subduction is occurring. Because the mountain range on the west coast of North America varies along its great extent, Lithoprobe sci-entists also did similar work across northern British Columbia, where, just west of the Queen Charlotte Islands, the largest recorded earthquake in Canada's history occurred. It's all a part of the most recent action explaining how North America came to look the way it does.

North America is being led on its ponderous journey west by a giant sea horse. The head, arcing curiously into the Pacific, is formed by the Brooks and Alaska mountains. The body comprises the Rocky Moun-tains, Coast Range and Cascades, and the belly the Sierra Nevadas. Through Mexico and Central America, the Sierra Madre ranges form the long narrow sweep of the tail. The sea horse is geologically young and still growing. Together, these mountain ranges form the North American Cordillera, similar to the ancient Appalachians and Grenville mountains, but there are significant differences, and to understand them we need to go back in time and a bit inland.

It's a truism of plate tectonics that yesterday's mountains are tomorrow's lowlands and this is nowhere more true than in Alberta and Montana.

When mountains rise, something invariably falls. Characteristically, deep basins filled with sediments occur along the flanks of rising mountain chains. If these basins grow offshore, they can end up squashed between two approaching continents and, themselves, become part of the growing mountain chain. However, basins also form where the crust buckles down behind mountains on the leading edge of a moving continent. There is a classic one behind the North American Cordillera, running from the Yukon to Mexico and reaching thicknesses of six kilometres. It is economically vital, containing hundreds of trillions of cubic feet of natural gas and tens of billions of barrels of oil, and it is refreshing to look at rocks that have not been cooked to a collection of swirling black and white bands. Of course, the older stuff is there too, just buried very deeply.

The ancient rocks, or basement, covered by the Western North American Sedimentary Basin is familiar—mind-bogglingly old microcontinents and superterranes welded together by vanished mountains. Old friends like Hearne and Rae are there, as are Wyoming and others, like Buffalo Head, that never see the light of day unless fragments are brought to the surface in incredibly deep drill holes. As we have seen, almost two billion years ago, Hearne-Rae-Slave and Superior came together in a titanic collision across the Trans-Hudson Orogen. It may or may not have been a part of a true supercontinent, Columbia, but it certainly was part of the stable core of North America, or Laurentia.

Over the past 1.5 billion years, eastern Laurentia has been an exciting place—Europe and Africa have come and gone, the Keweenawan and Iapetus oceans have opened and closed, and the Grenville and Appalachian mountains have risen and fallen as, perhaps one day, the Atlantic mountains will.

Of course, talking about the eastern and western edges of Laurentia is a bit misleading. That's what they are now, but Laurentia didn't keep still and what we comfortably call east and west now have probably faced every possible compass point. Over its long history, Laurentia has sat at the south pole for a while as part of Rodinia, drifted north to equatorial climes as a huge island home to countless doomed trilobites, stuck onto Europe and Africa to form Pangaea, and then again gone wandering across the new Atlantic. But, for the sake of argument, let's stick with east and west. So what was western Laurentia doing while the east was having all the fun?

Well, it wasn't exactly quiet, especially recently. Over the past 150 million years or so, sea floor equivalent to almost a third of Earth's circumference has been destroyed beneath Laurentia's relentlessly advancing western edge and one of the world's great mountain ranges, the North American Cordillera, has grown. But the history before that is a bit murkier.

Once upon a time, Siberia, Australia, and other fragments of what are today faraway places were attached to western Laurentia. Then, close to 1.8 billion years ago, the familiar story of rifting began. The crust stretched in the rifts, and deep basins began to fill with sediments washing off the nearby continents. In the northern Cordillera, Lithoprobe discovered one of these: the Fort Simpson basin, filled with sediments (now pressure-cooked) up to twenty-five kilometres thick. Similarly in the southern Cordillera, a depth of tens of kilometres of sediments, as old as 1.5 billion years and bereft of any fossil life, are preserved as the Purcell Supergroup (known as the Belt Supergroup in the United States). You drive between their towering walls if you cross the Purcell Mountains of southern British Columbia.

Seven hundred million years later, Laurentia was again involved in a supercontinent's demise. As Rodinia fragmented, basins formed off western Laurentia (which probably resembled the Atlantic coast of North America today), and filled with sediments that now form the Windermere and Gog rocks of the Rocky Mountains. The fossilized burrows of various busy creatures make their appearance in the late Precambrian Gog and are followed by the famous Burgess Shale fossils.

Through the Devonian and Jurassic, sediments continued to accumulate and carbonate reefs grew in the shallow, warm coastal waters. However, there were signs of approaching change. Within the deep water shales of the Devonian, around 380 million years ago, there are precursors of doom—thin layers of volcanic ash.

The ash came from a huge, curving arc of volcanic islands, similar to today's Aleutians, that lay offshore from Alaska to California. It wasn't a continent, and there were thousands of kilometres of open ocean to the west, but they did form a significant landmass. The arc formed above a subduction zone and about 365 million years ago it began to be shoved against Laurentia.

The immense forces involved had an effect as far inland as Alberta, Montana, and Nevada, pushing the sediments of the Purcell, Windermere, Gog, and later basins up onto the continent, gradually forming the imposing eastward-facing cliffs of the Rockies that attract so many tourists and calendar photographers.

As the pushing continued, the volcanic islands themselves reached the shore. The first pieces form the Klamath and northern Sierra Nevada Mountains of California and southern Oregon.

As the volcanoes continued to spew out lava and ash, the arc began collapsing against the continental edge. Now, the Atlantic began opening far to the east, dramatically increasing the forces at work as North America moved west. Bits and pieces of the island arc, with exotic names like Quesnellia and Stikinia, came ashore, welding onto Laurentia and pushing the Purcell and later sediments farther east. By around one hundred million years ago, the last major piece of the jigsaw, Wrangellia, met North America and the coast approached where it is now.

But things weren't finished. Sediments trapped between Wrangellia and the mainland were deformed, cooked, and pushed up. As the dinosaurs roared across the rising mountains, sediments filled in the smaller basins along the coast.

The movement of North America and the Pacific plate was not always at right angles. Around eighty-five million years ago, the Pacific Ocean floor began moving strongly northward. This smeared the pieces of the jigsaw along the continental edge and created huge faults. One runs along the Fraser Canyon where the west wall of the canyon has moved 130 kilometres north relative to the east wall. It also created the San Andreas Fault, which, as the residents of California know only too well, is still moving the southwestern piece of their state northward. Further north, the Queen Charlotte–Fairweather fault system, along which Canada's largest earthquake (magnitude 8.1) was generated, also is the product of the northward motion of the Pacific plate.

So a great mountain range formed, but it is different from the Grenvilles, the Appalachians, or the Alps. While Keweenawan and Iapetus opened and then closed again, the proto-Pacific Ocean kept opening and the continents that formed its western margin wandered away to have their own adventures on the surface of the ever-changing globe. The North American Cordillera did not form from two continents coming together, but rather from one scooping up many thousands of kilometres of oceanic crust along with whatever bits and pieces were sitting on it and, as Lithoprobe can show, this makes it fundamentally different.

In looking at jigsaw puzzles of Earth, we've graduated from a simple *Sesame Street* puzzle of the world, through a more complex collection of pieces that built Laurentia and the other continents, to the three-dimensional master's puzzle that is British Columbia west of a curving line drawn from about Nelson to Watson Lake, Yukon.

One of Lithoprobe's most striking achievements was the creation of a vertical cross-sectional map that runs from the Juan de Fuca ridge across this complete jigsaw puzzle. It shows not only where the great earthquakes happen, but all the slices that have piled up on this edge of Laurentia, and there are surprises.

The jigsaw puzzle of western North America consists of two hundred or so bits and pieces that have stuck onto or slid along the continental edge. Some are tiny (Alcatraz Island in San Francisco Bay is one) others are large (like the Brooks Range of mountains in Alaska). These pieces tell the story of what has happened here and how the continent has grown. One of the largest, most recent, and best-studied additions is Wrangellia, which stretches from east of Anchorage, Alaska, through the Queen Charlotte Islands and Vancouver Island, all the way to Hells Canyon in Idaho.

Wrangellia began 367 million years ago as volcanic and sedimentary island arc rocks, which were then covered by an immense outpouring of lava onto the ocean floor around 230 million years ago. It happened quickly (over about five million years) while Wrangellia sat somewhere near the equator. These basalts formed a broad oceanic plateau that probably formed extensive islands. As the islands subsided, shallow water, mud, and carbonates were deposited. One of the interesting things about these sediments is a small clam that lived happily in them. Daonella is a common fossil in Asia, but the rocks of Wrangellia are the only place it is found in North America.

Over fifty million years or so, Wrangellia drifted about four thousand kilometres north into cooler waters and into the path of North America. A little before one hundred million years ago, Wrangellia crashed into southwestern North America, perhaps as far south as Baja California. Then it began a long and strange journey, smearing itself up the coast all the way to Alaska.

Wrangellia is not unique, it is merely the latest of the pieces in the puzzle. Its story has been repeated many times and each time there was lateral, as well as back and forth, movement. We can tell all this from the surface geology, but what is going on deep below? Is it a simple continuation of what we see above or are there surprises? If Lithoprobe teaches one thing, it is that there are always surprises!

Logic, without the benefit of geological insight, would suggest that the ancient North American continent ends around where the prairies meet the mountains. As we have seen, examining the rocks shows that the Rockies are in fact huge slabs of ancient sediments that have slid on top of the continent. So North America proper, the Laurentia of 1.5 billion years ago, continues beneath the mountains, but how far?

Further study of the geology suggests that the continent ends along the line from Nelson to Watson Lake, where the jigsaw piece begins. But

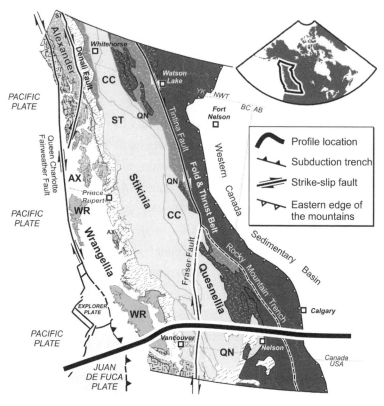

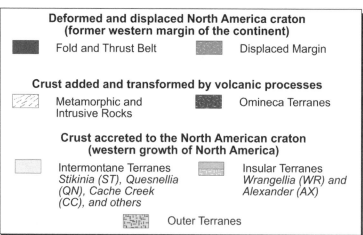

A geological map of British Columbia showing the main components of the crust. During the last two hundred million years, most of the crust west of the Tintina Fault–Rocky Mountain Trench has been added to western North America. The legend identifies three major types of crustal rocks; different geological processes were involved in generating the different types. Within the accreted crust, various terranes are identified. Some of the major faults are indicated. The thick solid line shows the location of the cross-sectional profile provided in the next illustration. The inset shows the location of the main map within North America.

Lithoprobe saw more. Deep beneath the puzzle west of Nelson, North America continues. It thins, but it is still there, all the way to the Fraser Canyon.

The situation is not as simple as first thought. North America is not a bulldozer crushing the ocean plate beneath it and piling islands and continental fragments along its edge. It has a prow, like an extended version of the ramming prows on Greek and Roman triremes, that slices through whatever it meets, forcing the lower bits down to a fiery fate in the mantle and smearing the top bits onto its upper edge. This slicing and smearing involves huge blocks of crust moving against each other and, as we have seen, this causes earthquakes, like the great San Francisco earthquake of 1906. However, the only part that remains active today is the west coast region. Inland, the many blocks of crust have been more or less quiescent for the last forty million years. But all this action over tens of millions of years has caused other effects.

Heating associated with subduction and collisions, aided by the heat generated by natural radioactivity in crustal rocks, causes partial melting of rocks in the deep crust and mantle. The buoyant melted rocks force their way upwards along the subduction zone, forming the chain of volcanoes like Mount Garibaldi and Mount St. Helen's along the west coast. More inland, they squeeze their way nearer the surface where they slowly cool, forming large areas of granite rocks. Mother Earth is a complex old lady.

## Profile across the Southern Canadian Cordillera

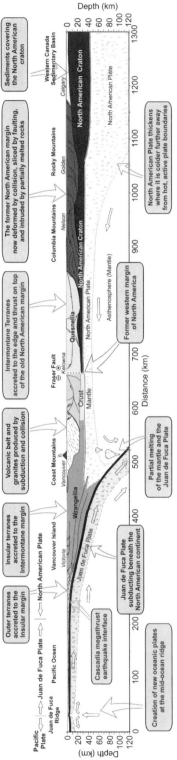

A lithospheric cross-section from the surface to a depth of 120 km across southern British Columbia; the profile location is given in the previous illustration. The section is based on seismic studies coupled with many geological and other geophysical studies. Explanations for the various parts of the cross-section are given by text boxes with arrows. Some major features and cities are indicated above the section. Thin lines within the crust indicate interpretations of the complex structures that form it. Open arrows within the section indicate current motions that are taking place, mainly in the mantle. Note that rocks of the North American craton extend as a thinning wedge at the base of the crust all the way to the Fraser Canyon. The rest of the crust is built upon this ancient foundation. The asthenosphere (white with no pattern) is the partially molten part of the upper mantle upon which the lithosphere rests. The open arrows in it indicate the upper part of a mantle convection cell.

# Fishing from a Greyhound
by John Wilson

FIELD WORK RARELY PRESENTS MUCH TIME or opportunity for anything other than the most basic recreational pursuits. Going out for
dinner and a movie is difficult if the nearest restaurant and movie theatre
are a helicopter ride away. There are exceptions, though.

Back in the late 1970s, I was working on coal exploration in northern
BC. It was magnificent country and the work, with the exception of being
deafened in a helicopter for several hours each day, was wonderful.
There was even time for fishing and, when the work was done, two tired
but happy geologists lugged black garbage bags full of ice and spring
salmon onto the *Grumman Goose* from Stewart to Vancouver. That was
when things began to go wrong.

We were based in Calgary and had bookings on a flight across the
mountains that afternoon. Unfortunately, our airline was grounded by
a strike. What was flying out of Vancouver was packed. We put ourselves
on wait lists and sat around, nervously watching our bags of salmon.

After three flights had left, we had moved from thirty-third to thirty-
first on the wait list. Obviously, we would be cooking and eating the
salmon in the departure lounge before we got a seat. In desperation, we
took a cab to the Greyhound bus depot and booked seats on the seventeen-
hour overnight bus to Calgary.

All very well, except the salmon had been in the garbage bags for
some time already and the ice was now cool water. We emptied the water
out and rushed over the road to a nearby motel where we raided the ice
machines. The understanding driver let us put the bags in the belly of
the bus and we settled in.

It was summer, it was hot and I couldn't sleep. Partly because I don't
sleep well on buses, but mostly because, with every kilometre, I expected

the smell of fish to waft up from below my feet.

What saved us the embarrassment of rotting fish was that Greyhound buses have to stop every couple of hours for the driver to take a rest. At the first stop we leaped out and tore around finding ice to replenish the bags.

People were curious—well, not all of them. Some just shook their heads in pity, and others looked at us as if it was a particularly cruel fate that had placed us on their bus. As the story of the fish spread through the bus, we acquired helpers. By the time we stopped at Golden, people were leaping down the instant the doors opened and fanning out in all directions in search of fresh ice. I like to think it was human kindness, but I suspect no one was keen on spending the last five hours on a bus with kilos of rotting fish.

Anyway, we and the fish made it safely to Calgary. The barbecue was awesome!

Courtesy of John Wilson

*Salmon prior to its road trip to Calgary.*

# PART 5

# OLD AGE:
# THE NEXT BILLION YEARS

CHAPTER 8

# A QUIET OLD AGE? NOT LIKELY.

Study the past if you would define the future.
—Confucius

Telling the future by looking at the past assumes that
conditions remain constant. This is like driving a car by
looking in the rear-view mirror.
—Herb Brody

Prediction is very difficult, especially about the future.
—Niels Bohr, Danish physicist

FOR SOME, PREDICTING THE FUTURE IS EASY. All you need do is read
through the quatrains of Nostradamus and all becomes clear. Unfortunately, the downside is that Nostradamus's verses only seem to be understandable after an event, when it is possible to squeeze meaning out
of the French mystic's vague language. I suspect that Bob Dylan's "Des-
olation Row," approached in the correct state of mind, is equally effective
in the prediction department. It's certainly more tuneful.

Some aspects of predicting the future are as easy as Nostradamus
seemed to find them—at least in the short term. Tomorrow the world

will continue to turn, the Sun will come up and North America will move another fraction of a millimetre farther from Europe. In the longer term, some things will continue on their inevitable, predictable course: Earth's core will continue to generate heat, the Sun will burn more fuel, and the Moon will move ever farther from us. The consequences of those things can be guessed at—eventually the heat from Earth's core will be insufficient to drive plate tectonics, the Moon's gravity will not pull the tides, and the Sun will burn out and die. The trick is to know when, and what smaller-scale wrinkles will occur on the way. We know, broadly, how plate tectonics works, so it is reasonable to assume that it will continue to do so. However, since we don't fully understand the mechanism that drives it, we cannot say what will happen in the longer term, or even what variation there might be in the shorter term.

Earth is an immensely complex system with countless factors at play, many of which we either only vaguely understand or have not yet guessed. Change one factor and the others all move about to rebalance the whole. Given that our current state of knowledge, impressive though we like to think it is, ignores a whole bunch of these factors that we cannot even imagine, understanding anything but the simplest process we see around us seems to be an exercise in frustration and perhaps futility.

A problem with interpreting the past, as with predicting the future, is that we can only see what we can imagine. It's getting back to guessing what Arthur Holmes' alarm mechanism might be for. We can only imagine what we have seen or know, and from that we have a penchant for building theoretical edifices on the scantiest information. We think we know about the evolution of life, but 99.9 per cent of everything that ever lived never made it to the fossil record and is, therefore, unknowable.

Take a mundane example. You have misplaced your house keys. Normally, you put them in that bowl by the front door, or in your jacket pocket, or on the bedside table, but they are in none of those places, so you expand the search. You can imagine them falling off the bedside table, so you look on the floor. You even look behind the end of the bed, despite the fact that the keys would probably need legs to get there. You check the pockets of other jackets, even the one you haven't worn since you went skiing a fortnight ago. And you empty out the vase of dried flowers in the front room in case you had a fit, put the keys in it and then suffered a very specific amnesia.

In short, you go to extraordinary lengths to rationalize what you expect to be the case. You take what you "know" to extremes while ignoring the fact that the keys are actually sitting in plain sight on the (admittedly cluttered) kitchen table. You walked past them a dozen times, even scanned the tabletop, but you didn't see the keys because you don't ever put them there, so you can't imagine them there, and therefore you don't see them.

The lost-keys scenario has happened countless times in the history of science. A theory, such as any one of the pre-radioactive methods for determining the age of Earth, is entrenched in most people's minds. Intelligent people will go to sometimes extraordinary lengths to "prove" their favourite theory. Bishop Ussher believed the Bible explained it all and so spent years adding up the ages of Old Testament prophets to come up with a ridiculously precise figure. Lord Kelvin did the same thing, although he based his beliefs on the science he had available to him at the time.

In the same way, the simple, overall beauty of plate tectonic theory blinds us to things that might not fit. Subduction zones and spreading ridges have revolutionized the way we think of our planet. They explain a dynamic Earth, and give us a framework to look at everything from the deepest ocean trench to the highest mountain ridge in a new light. But there's a downside. Plate tectonics focuses our view on the active bits, oceanic ridges where plates form, or edges where they are destroyed or grow, and that is only a tiny fraction of the planet's surface. Despite maps that show everything in convincing shading and colours, the deep ocean floor, almost four-fifths of our home's surface, is mapped in less detail than the surface of Mars. Certainly there is an enormous amount of stuff out there that will surprise us when we find it. Some of it will require a re-evaluation of our explanation of plate tectonics. Even in our explanation of the land masses, the fraction that we know best, there are still many things that don't fit.

Except for the leading edge crumpling its way over the Pacific floor, North America should be a quiet place. Big earthquakes and volcanoes only happen at plate edges or when a supercontinent breaks apart. Los Angeles, San Francisco, Seattle and Vancouver are places where everyone should have earthquake insurance—but Kansas! Apart from tornadoes and a strange link to the land of Oz, Kansas should be a safe place, but it sits right between two of the most dangerous places in the world—neither of which can be completely explained by plate tectonics.

If three earthquakes the size of the ones that devastated New Madrid in the winter of 1811–12 were to strike now where Kentucky, Tennessee, Arkansas, Missouri, and Illinois all come together, it would be a tragedy of extraordinary proportions. The shaking would be felt almost all over the continent, thousands of people would die, hundreds of thousands would be left homeless, and Memphis would be largely flattened.

We know that the New Madrid quakes occurred on the Reelfoot Rift system, probably along a failed arm of the Iapetus Ocean, but why is it still shaking five hundred million years after its chance at glory? Plate tectonics does not have the answer, nor can it predict when the crust beneath New Madrid will shudder again.

Just as there are megathrust earthquakes, there are super-volcanoes. Fortunately, humankind hasn't seen one of these. But if we do, it could well be the last thing many of us do see.

The sites of several past super-volcano eruptions have been recognized. Some are where you would expect them to be, near active plate boundaries in Indonesia, Japan, and New Zealand, but some are not. One of the largest is visited by three million people a year.

Yellowstone Park last blew up 640,000 years ago with the force of 2,500 Mount St. Helenses, or roughly the equivalent of the impact of a one-kilometre-diameter asteroid. It created a caldera (a disproportionately large crater) eighty-five kilometres long by forty-five kilometres

*Are these delicate terraces in Yellowstone Park where the next super-volcano will erupt?*

wide and spewed out enough ash to cover the United States to a depth of twelve centimetres. It devastated much of North America and degraded the global climate for many years. According to Professor Stephen Self of Britain's Open University, such a super-eruption now could "result in the devastation of world agriculture, severe disruption of food supplies, and mass starvation." To round out the cheerful picture, Professor Self added, "These effects could be sufficiently severe to threaten the fabric of civilization."

When will this happen? There have been three super-eruptions at Yellowstone with an average interval of six hundred thousand years, so we were due for the next one forty thousand years ago. Of course, it might not happen for tens of thousands of years yet, or it may never happen, although it would probably not be wise to put a long-term wager against the latter. Apart from probably having trouble picking up your winnings, parts of Yellowstone have risen by seventy centimetres in the past century, suggesting that something is going on below Old Faithful.

Geologists assume that the Yellowstone super-volcano was caused by a hot spot below, a plume of hot molten material rising from the core and pooling below the crust until the pressure builds to breaking point. But there is a lot of guesswork involved and no one knows how hot spots relate to plate tectonics or the still largely mysterious mechanisms that drive it. At least we will know why it's happening, when Los Angeles and Vancouver topple to the ground in earthquakes.

So a lot of things can, and will, happen that we don't understand and cannot predict. This is hardly surprising. We cannot predict much about systems that are operating at present and are relatively easy to study— they're just too complicated. Take global warming as an example. We know greenhouse gases are collecting, that the ozone layer is being damaged and that the global temperature is increasing. Reasonably, this leads to the assumption that ice caps will melt and sea levels will rise. Unfortunately, this tells us nothing about what the weather will be like next summer in your backyard, or even how it will affect major ocean currents like the Gulf Stream and, consequently, the entire agricultural potential of Europe. Think how much more difficult it is to interpret information from the past that is incomplete and unexpected, and possibly arises from processes that are unknown. Then try to extrapolate that into an unknown future. Realistically, the best we can do is tip our hats to the uniformitarians, cross our fingers, and hope for the best. But speculation is fun, so here goes. Let's begin by going back a bit.

One of the oversimplifications of plate tectonics and the supercontinent cycle is the impression it gives of distinct phases of continental breakup and reassembly. You could be forgiven, with all the talk of Pangaea breaking apart, the Atlantic opening and North America ploughing its inevitable way across the Pacific, for thinking that we are in a break-up phase right now. In fact, most of the sixty-five million years since we bade farewell to the dinosaurs have been characterized more by plate collisions than separations.

The last major plate tectonic rifting took place around fifty million years ago when Australia left Antarctica. Since then, the only rifting has been minor: Arabia moving away from Africa, Japan drifting into the Pacific and California heading north. Almost immediately after Australia began its long journey north, the first big bump occurred.

When India hit Asia, it was travelling at a speed of around fifteen centimetres per year, possibly a modern plate tectonic record. India was old and solid, whereas Asia had only recently knitted together from other fragments. As India snowplowed in, the joints in Asia reactivated and the bits were squashed up and to the side, which must have caused some pretty impressive earthquakes to disturb our mammal ancestors.

This was the beginning of the end of the Tethys Ocean and the start of an impressive collection of mountain ranges. Africa was heading for Europe, pushing bits and pieces in front of it. Turkey, Greece and the Balkans created the mountain ranges that make southeastern Europe so rugged. The Alps rose in front of Italy, and the Pyrenees bunched up between Spain and France. As the Sumatra-Andaman earthquake so devastatingly showed, Australia is colliding with Indonesia and southeast Asia.

So that's where we are, bumping and grinding our way into the future. What will it look like? In the short term, geologically speaking, what we see now will continue. Australia will crash into southeast Asia, crushing all the bits and pieces of volcanic arc and continent on the way. Africa will recreate the experience of India colliding with Asia. As it hits Europe, the Mediterranean Sea will close, as will the Red Sea. Eventually, perhaps fifty million years down the road, a Himalayan-scale mountain chain will run from Spain through southern Europe and the Middle East to link with the mountains already risen in Asia. This will be one of history's great mountain ranges, on a par with the mighty Grenvilles of a billion years earlier.

But not all the activity is straightforward collision. There is rotation going on as well. Fifty million yeas down the road, Europe/Asia/Africa will have rotated clockwise, bringing Western Europe closer to the North Pole and Siberia closer to the Equator. Similarly, North America will rotate in the opposite direction. In Africa, a new sea will open up as East Africa separates from the main African continent.

This is all fairly straightforward, but what happens next is not. Continuing as things are, North and South America should continue their adventurous journey, destroy the Pacific Ocean and, eventually, crash into Asia to form the next supercontinent. Unfortunately, that doesn't seem to be the way things usually happen. Iapetus didn't keep growing. It reached a respectable size and then closed again. This would suggest that the Atlantic will one day stop widening, a subduction zone will form along the eastern seaboard of the Americas, and we will begin approaching Europe.

Why?

Who knows! Perhaps it will be caused by something to do with the mysterious mechanisms that drive convection in the mantle. Let's assume that not all the convection plumes rising from Earth's core have the same strength. Let's say that there is a greater heat flow beneath the Pacific than beneath the Atlantic. In fact, let's say that the convection cell that is forcing the Atlantic open is only there because of a huge heat build up beneath that thick piece of continental crust, Pangaea.

As the Atlantic widens, the continental crust above the plume is replaced by oceanic crust. This is much thinner, allowing heat to escape more readily. Therefore, as the Atlantic widens, the convection cell that is forcing it open weakens as it loses heat. Eventually, the forces driving North America westward are not strong enough to overcome the forces beneath the Pacific pushing in the opposite direction. Then, America will stop its travels, subduction will begin from Labrador to Tierra del Fuego, mountains will rise in New England and Brazil and the Atlantic will close.

Maybe.

Let's say it does happen this way. One hundred and fifty million years from now, the Atlantic will be about half today's width and the mid-Atlantic spreading ridge will be largely subducted beneath the mountains where New York and São Paolo once stood. Europe/Asia/Africa/Australia, complete with its impressive mountain range along the middle, will continue to rotate, forcing Australia back against Antarctica (another ocean opening and closing).

A hundred million years later, the Atlantic will be no more. North America will face southern Africa across a mighty mountain range. South America will curl around Africa, its tip tucked in beside northern Australia and Indonesia, although these terms will long ago have lost any meaning. We will have a new supercontinent, perhaps covered with searing deserts as Pangaea was in the Permian, or perhaps freezing towards another snowball Earth. Then it will all begin again, heat will build up beneath the new continent, rifts will form, and oceans will open. A billion years from now, Earth might be back looking something like it does now. But mountains, now ghosts, will have risen, and oceans will have vanished.

Of course, this is all wildly speculative. The only thing we can be certain of is that, until the energy generated by radioactivity has halved and halved and halved itself to almost nothing and the furnace in the Earth's core flickers and dies out, bringing everything to a halt, things will continue to happen. Plates will form, move, and die in an ever-changing dance. And through it all, life will find a way, sometimes encouraged by the moving crust beneath it and sometimes drawing on its most fundamental resources just to survive. We will not be around to see all this, but something probably will. Until then, all we can do is live in the instant of time we have been given and struggle, through Lithoprobe and our capacity for imagination, to understand a tiny fraction more of this strange place we call home.

# ACKNOWLEDGEMENTS

W E'D LIKE TO THANK a number of people for their contribution to this book. Phil Hammer of Earth and Ocean Sciences at the University of British Columbia and Doug Hunter generated the illustrations. Senior editor Michael Mouland and associate editor Carol Harrison at Key Porter Books ably guided its preparation. About a thousand earth scientists contributed to Lithoprobe's research from 1984 to 2005. Their efforts led to the project's international acclaim and to results included herein. Continuous funding for Lithoprobe derived from the Natural Sciences and Engineering Research Council (NSERC) of Canada and the Geological Survey of Canada. Provincial and territorial geological surveys contributed when Lithoprobe studies were in their jurisdiction; industry contributed when the studies aligned with their interests. NSERC also provided financial support for Ron Clowes's research activities, including preparation of this book.

# NOTES

1.  D.J. McLaren, "Earth Science and federal issues," *Geoscience Canada* 8 (1981): 106.

2.  Ibid., 111.

3.  International Lithosphere Program's website, http://sclilp.gfz-potsdam.de/ (accessed July 17, 2008).

4.  www.unido.org/userfiles/PuffK/L_pollutants.pdf

5.  John Ross, *Narrative of a Second Voyage in the Victory, in Search of the North-west passage* (1835). Cited in M.J. Ross, *Polar Pioneers* (Montreal: McGill-Queen's University Press, 1994).

6.  James Clark Ross, *Voyage of Discovery and Research in the Southern and Antarctic Regions during the years 1839–43* (1847).

7.  Internet Sacred Text Archive, "The Third Tablet," http://www.sacred-texts.com/ane/stc/stc06.htm (accessed July 17, 2008).

8.  Camille Flammarion, *Omega: The Last Days of the World* (Lincoln, NE: University of Nebraska Press, 1999).

9.  Mark Eppler, *The Wright Way* (New York: AMACOM Books, 2007). Excerpt from chapter 1 on publisher's website, www.amanet.org/books/catalog/0814407978_complexity.htm (accessed July 17, 2008).

10. Zimmer, "Geophysics: Ancient Continent Opens Window on the Early Earth," *Science*, December 17, 1999, 286, no. 5448, 2254–2256. Available online at www.sciencemag.org/cgi/content/full/sci;286/5448/2254 (accessed July 17, 2008).

11. T. H. Huxley, *Man's Place in Nature* (New York: Modern Library, 2001).

12. Government of Nova Scotia, "Titanic: The Unsinkable Ship and

Halifax," http://titanic.gov.ns.ca/wireless.html (accessed July 17, 2008).

13. Geological Survey of Canada, "Elkanah Billings: the first Canadian paleontologist," in *Past lives: Chronicles of Canadian Paleontology*, http://gsc.nrcan.gc.ca/paleochron/18_e.php (accessed July 17, 2008).

14. Brian F. Atwater, Musumi-Rokkaku Satoko, Satake Kenji, Tsuji Yoshinobu, Ueda Kazue, and David K. Yamaguchi, "The Orphan Tsunami of 1700–Japanese Clues to a Parent Earthquake in North America," U.S. Geological Survey Professional Paper 1707 (2005), published with University of Washington Press. Available online at http://pubs.usgs.gov/pp/pp1707/ (accessed July 17, 2008).

# INDEX